LANGE

USMLE ROAD MAP

HISTOLOGY

HAROLD J. SHEEDLO, PhD

Assistant Professor
Department of Cell Biology and Genetics
University of North Texas Health Science Center
Fort Worth, Texas

Lange Medical Books/McGraw-Hill
Medical Publishing Division

New York   Chicago   San Francisco   Lisbon   London   Madrid   Mexico City
Milan   New Delhi   San Juan   Seoul   Singapore   Sydney   Toronto

USMLE Road Map: Histology

4 5 6 7 8 9 0 DOC/DOC 0

ISBN: 0-07-144012-7
ISSN: 1553-6769

Notice

This book was set in Adobe Garamond by Pine Tree Composition, Inc.
The editors were Janet Foltin, Harriet Lebowitz, and Karen W. Davis.
The production supervisor was Richard C. Ruzycka.
The illustration manager was Charissa Baker.
The illustrations were rendered by the Dragonfly Media Group.
The designer was Eve Siegel.
The index was prepared by Andover Publishing Services.
RR Donnelley was the printer and binder.

This book is printed on acid-free paper.

CONTENTS

USING THE
USMLE ROAD MAP SERIES
FOR SUCCESSFUL REVIEW

What is the Road Map Series?
Short of having your own personal tutor, the USMLE Road Map Series is the best source for efficient review of major concepts and information in the medical sciences.

Why Do You Need A Road Map?
It allows you to navigate quickly and easily through your histology course notes and textbook and prepares you for USMLE and course examinations.

How Does the Road Map Series Work?
Outline Form: Connects the facts in a conceptual framework so that you understand the ideas and retain the information.

Color and Boldface: Highlight words and phrases that trigger quick retrieval of concepts and facts.

Clear Explanations: Are fine-tuned by years of student interaction. The material is written by authors selected for their excellence in teaching and their experience in preparing students for board examinations.

Illustrations: Provide the vivid impressions that facilitate comprehension and recall.

 Clinical Correlations: Link all topics to their clinical applications, promoting fuller understanding and memory retention.

 Clinical Problems: Give you valuable practice for the clinical vignette-based USMLE questions.

 Explanations of Answers: Are learning tools that allow you to pinpoint your strengths and weaknesses.

To my mother (Jill), father (Harold), brothers (Gary, Darryl, Robin, Terry, Steve), and sister (Susan).

Acknowledgments

Thanks are extended to Robert J. Wordinger, PhD, and Rustin E. Reeves, PhD, for resourses used to generate histologic images for this project and D. Maneesh Kumar, PhD/DO student (Department of Cell Biology and Genetics), for his valuable critique.

CHAPTER 1
CELL BIOLOGY

I. Plasma Membrane

A. General features

1. The plasma membrane, also called the **plasmalemma,** separates the cell from its environment and encloses the intracellular compartments, nucleus and cytoplasm, within a cell.
2. It is formed primarily of a **phospholipid bilayer,** consisting of phospholipids, glycolipids, and cholesterol and peripheral and integral proteins.
3. Glycoproteins and glycolipids on the surface of a membrane constitute its **glycocalyx.**

B. Membrane events

1. **Pinocytosis,** a form of **endocytosis,** is an uptake of extracellular small material, such as ions, by small vesicles.
2. **Phagocytosis,** a form of **endocytosis,** is an uptake of large particles, such as bacteria and cellular debris.
3. **Exocytosis** is the fusion of a cellular vesicle with the plasma membrane and release of its contents into the extracellular space.

II. Nucleus

A. The **nucleus** is the center of cellular activity and contains chromosomes and nucleoli (Figure 1–1A).

B. The **nuclear envelope,** which forms a barrier between the nucleus and the cytoplasm, contains **pores** for exchange of macromolecules to and from the cytoplasm.

1. The nuclear envelope has a structure common to all eukaryotic cells but is modified in some cells by variations in size, number, structure, and disposition of its components.
2. The envelope is a **tripartite structure** composed of inner and outer nuclear membranes, separated by a clear space, called a **perinuclear cisterna.** Total nuclear membrane thickness is 30 nm.
3. The nuclear **lamina** is a lattice structure of specialized intermediate filaments.
4. Phosphorylation of nuclear **lamins** results in disassembly of the nuclear lamina and vesiculation of the nucleus during **prophase.**
5. Reassembly of the nuclear envelope requires removal of phosphate residues from lamins during **late anaphase** or **telophase.**

C. **Nuclear pores** are formed at the site of fusion of 2 nuclear membranes (Figure 1–1B). These pores act as semipermeable sites in the nuclear membrane.

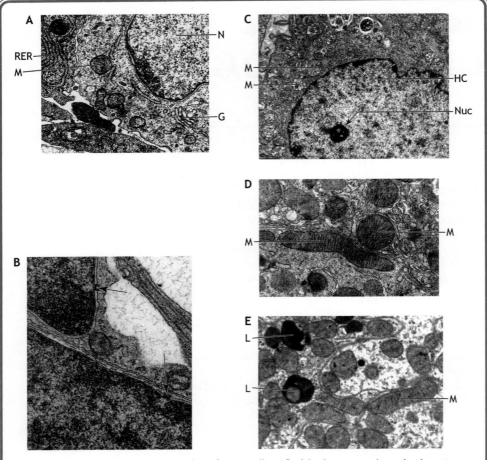

Figure 1–1. Electron micrographs of organelles. **A:** Nucleus, rough endoplasmic reticulum, and Golgi apparatus. **B:** Nuclear pores (arrows). **C:** Nucleolus. **D:** Mitochondria. **E:** Lysosomes. (N = nucleus; M = mitochondrion; RER = rough endoplasmic reticulum; G = Golgi apparatus; Nuc = nucleolus; HC = heterochromatin; L = lysosome.)

1. The size of nuclear pores varies from cell to cell, averaging 40–100 nm.
2. The number of nuclear pores varies depending on the age, metabolic state, and stage of differentiation of the cell.
3. The nuclear pore is filled with a diffuse substance, resembling a **diaphragm,** and in some cells consists of 8 subunits.

III. DNA and the Chromosome

A. Chromosomes

1. **Chromosomes** are formed by condensation of chromatin before nuclear division.

    2. In eukaryotic cells, each chromosome consists of 2 **chromatids.** A human somatic cell has 46 chromosomes and 4N (copies) DNA, called **diploid.**

    3. Chromatids are connected at a **centromere** that is a specific sequence of **deoxyribonucleic acid** (DNA) required for chromosome segregation.

B. Heterochromatin

    1. Heterochromatin is the region of a chromosome that remains condensed during interphase and early prophase.

    2. Ribonucleic acid (RNA) synthesis does not occur in these regions; thus, they are generally transcriptionally inactive.

    3. One of the 2 X chromosomes in many cells in females is condensed in interphase. This structure is called **sex heterochromatin** or **Barr body** and is often associated with the nuclear envelope.

    4. Early in development, 1 X chromosome of a female is **randomly** inactivated, remaining as heterochromatin. Thus, only 1 X chromosome is transcribed. The process of **random X chromosome inactivation** is largely unknown but may involve an RNA coat that induces heterochromatin.

    5. In differentiated mature cells, such as neurons, much of the DNA is in the **heterochromatic** state.

C. Euchromatin

    1. Euchromatin is the state in which the chromosome is unfolded or uncoiled to its maximum level and transcriptionally active (RNA synthesis).

    2. Euchromatin is typical of undifferentiated immature cells, such as **stem** or **progenitor cells.**

    3. Ribosomal RNA (rRNA) is the key element of ribosomes and is synthesized by RNA polymerase I. **Messenger RNA** (mRNA) is the template for protein synthesis and is synthesized by RNA polymerase II. **Transfer RNA** (tRNA) transports a specific amino acid to the site of protein synthesis and is synthesized by RNA polymerase III.

D. Histone proteins

    1. Histone proteins are designated **H1, H2A, H2B, H3, and H4,** and lysine and arginine account for about 25% of each histone protein. These basic amino acids provide histones with a positive charge.

    2. An **octamer of 2 of each histone proteins** wraps around the negatively charged DNA molecule twice to form a **nucleosome,** which consists of approximately 200 base pairs.

    3. H1 of 1 nucleosome binds to an H1 of an adjacent nucleosome to package chromatin forming a 30-nm fiber.

    4. Once thought to be involved only in the maintenance of DNA structure, **histones** are now also known to play a role in gene regulation and may also inhibit transcription.

E. Karyotype

    1. A **karyotype** is a representation of chromosomes during **metaphase,** arranged in sets of 2 from 1 to 22 with an X and an X or Y chromosome, designated 46,XY or 46,XX.

    2. Chromosomes are arranged in **decreasing order of length:** 1–3 (group A), 4–5 (group B), 6–12 and X (group C), 13–15 (group D), 16–18 (group E), 19–20 (group F), and 21–22 and Y (group G).

3. Factors used to distinguish chromosomes to determine a karyotype include relative chromosomal size, position of centromere (region of attachment of 2 chromatids), length of individual chromosome arms (p = short arm, q = long arm), and banding patterns after Giemsa staining.

4. Chromosomal anomalies and the associated diseases are described in Table 1–1.

F. **Nucleolus**

1. The **nucleolus** is a region of the nucleus that is specialized for the production of **rRNA** and consists of a filamentous network made up of nucleoprotein fibers (Figure 1–1C).

2. Within the nucleolus are chromosomes containing loops of DNA and large clusters of rRNA genes.

Table 1–1. Karyotypic anomalies.

Disorder	Karyotype	Incidence	Clinical Signs
Cri du chat syndrome	46,XX(Y), 5p-		Mental retardation Cardiac anomalies
Down syndrome[a] (mongolism)	47,XX(Y), +21	1 in 700	Heart disease Leukemia Neurodegeneration Immune diseases
Edwards' syndrome	47,XX(Y), +18	1 in 8000	Similar to trisomy 21
Fragile X syndrome	46,XXq27.3	1 in 2000 (female) 1 in 1000 (male)	Mental retardation Large mandible
Klinefelter's syndrome[b]	47,XXY(Y)	1 in 850	Reduce fertility Skeletal abnormalities Atrophic testes
Patau's syndrome	47,XX(Y), +13	1 in 15,000	Similar to trisomy 21
Turner's syndrome[c]	45,X	1 in 3000	Abnormal secondary sex characteristics Short stature Amenorrhea

[a]Results from nonjunction of chromosomes during meiosis and is the most common chromosomal disorder.
[b]Results from nondisjunction during meiosis of one of the parents.
[c]Results from monosomy of the X chromosome and is the most common sex chromosome disorder in females.

3. Each gene cluster is known as a **nucleolar-organizer** and is the region of DNA bases that codes for rRNA. These structures are present on chromosomes 13, 14, 15, 21, and 22.
4. The nucleolus consists of **pars granulosa,** central dense granular material, and **pars fibrosa,** peripheral thin filaments.
5. The size and shape of nucleoli are dependent on their **activity** and are generally visible only during interphase.

IV. Cell Cycle

A. The cell cycle is controlled by **cyclins B, D, and G$_1$** and **cyclin-dependent protein kinases (Cdc2)** (Figure 1–2).

B. Activation of the cyclin B-Cdc2 complex, also called the **M phase-promoting factor,** allows the cell to enter the mitotic phase.

C. The **M phase** is the series of events during mitosis.

D. No DNA replication occurs during the **G$_1$ (G = gap) phase;** therefore, it is part of interphase.
1. The G$_1$ phase lasts from 3–4 days depending on such factors as age, species, and cell state.
2. Differentiated cells such as neurons are in the **G$_0$ phase.**

E. Formation of the cyclin G$_1$-Cdc2 or cyclin D-Cdc2 complex triggers entry of the cell into the **S (synthesizing) phase.**
1. **DNA replication** occurs during the S phase.
2. The resulting nucleus is tetraploid; thus, the cell has a diploid number of chromosomes, and each chromosome consists of 2 chromatids.

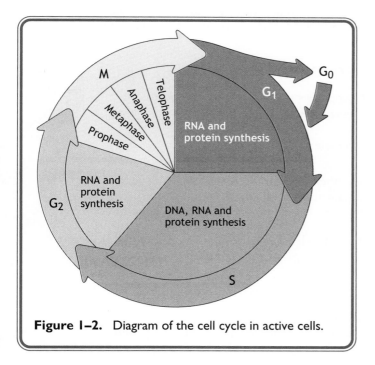

Figure 1–2. Diagram of the cell cycle in active cells.

3. The S phase lasts from 7–8 h, but its length is not dependent on cell conditions once the mitotic event has been triggered.

F. The **G₂ phase,** after DNA synthesis, usually lasts 2–5 h and occurs before the cell undergoes another nuclear and cell division.

p53 GENE MUTATIONS AND TUMORS

- *The gene product of the **p53** gene causes G₁ arrest and subsequently induces either repair of DNA or programmed cell death, specifically via **apoptosis.***
- *Loss of the p53 gene leads to such cancers as lung, breast, and colon carcinomas.*
- *More than 50% of tumors have a mutation of the p53 gene.*

V. Cell Division and Cytokinesis

A. **Mitosis**
  1. During **interphase,** chromatin is maximally uncoiled so that chromosomes are not distinguishable.
  2. During **prophase,** the mitotic apparatus is assembled, and the chromosomes become visible and make their first movements. The chromosomes appear as slender threads but become progressively thicker and shorter during this phase. At the end of this phase, the nuclear envelope begins to break down.
  3. During **metaphase,** the chromosomes, consisting of 2 chromatids, are arranged along the equatorial plate. **Microtubules** of the mitotic spindle are attached to the kinetochore.
  4. During **anaphase,** sister chromatids are pulled to opposite poles by action of the microtubules.
  5. During **telophase,** the chromatids have completed their migration, and the nuclear envelope begins to reform. A cleavage furrow develops at the equator of the cell.

RETINOBLASTOMA GENE MUTATIONS AND OSTEOSARCOMA

- *In the homozygous state, the **retinoblastoma (rb)** gene prevents unregulated cell proliferation.*
- *If both copies of the rb gene are mutated, tumor cells arise from retinoblasts in both eyes of children usually before 5 years of age.*
- *Patients with an rb mutation have a 100-fold greater risk of developing **osteosarcoma.***

B. **Cytokinesis**
  1. **Cytokinesis** results in a division of the cytoplasm, forming 2 daughter cells.
  2. This division can occur from mid-anaphase to telophase.
  3. During this process, the dividing cell lengthens, and the central zone becomes progressively narrower.
  4. This process leads to the formation of a **cleavage furrow,** which deepens, ultimately resulting in the fusion of the 2 separate plasma membranes and formation of 2 daughter cells.

VI. Endoplasmic Reticulum

A. **Rough Endoplasmic Reticulum (RER)**
  1. **RER** is a network of flattened sacs or **cisternae** that are studded with **ribosomes** (Figure 1–1A).

2. Most of the proteins synthesized by the RER are transported via vesicles to the Golgi apparatus, where they are glycosylated or otherwise modified.

B. Smooth Endoplasmic Reticulum (SER)
1. **SER** consists of a series of cisternae or flattened sacs that lack ribosomes usually found near lipids or glycogen within the cytoplasm.
2. The functions of SER include synthesis of **glycogen** and **steroids,** drug detoxification, and transport of electric potentials by a specialized SER called sarcoplasmic reticulum in striated muscle.

VII. Ribosomes

A. **Ribosomes** are found free in the cytoplasm, form polysomes, or are attached to endoplasmic reticulum (RER).

B. These organelles are spherical particles 15–25 nm in diameter that consist equally of RNA and protein.

C. In eukaryotic cells, ribosomes consist of a **60S subunit** and a **40S subunit.**

D. These organelles serve as the site of protein synthesis for either secretion (**RER**) or intracellular use (**polysomes**).

VIII. Golgi Apparatus

A. The **Golgi apparatus** consists of flattened plates of 2–12 cisternae, usually found near the nucleus (Figure 1–1A).

B. The *cis* face of the Golgi apparatus is the site of integration of vesicles from the RER, whereas the *trans* face is the site from which vesicles exit as lysosomes or as membrane-bound products that are secreted.

C. In addition to secretion, the Golgi apparatus functions to transport proteins embedded in vesicles into the cell membrane, such as **receptors** and **glycosylation** and **sulfation** of proteins.

CYSTIC FIBROSIS

- *The **cystic fibrosis** gene, found on chromosome 7, codes for a chloride ion channel protein **CFTR** (cystic fibrosis transmembrane conductance regulator).*
- *It is the most lethal genetic disorder in Caucasians, affecting 1 in 1500 to 1 in 4000 live births.*
- *The most common mutation of the CFTR protein results in defective processing of the protein from the **RER** to the **Golgi apparatus.** The protein is not properly folded or glycosylated and becomes degraded within the cytoplasm.*
- *Mutation of the CFTR protein leads to increased chloride in sweat and decreased water in the mucus.*
- *These patients suffer from mild to severe symptoms, including cirrhosis, lung disease, pancreatic disorders, and absence of male genital tubules.*

IX. Mitochondria

A. **Mitochondria** are double-membrane organelles that are ovoid or spherical in shape. The outer and inner membranes create an intermembrane space (Figure 1–1D).

B. These organelles are 0.5–1 μm wide and >5 μm long.

C. The inner membrane projects into the matrix of the mitochondria to form **cristae.** These structures have a vesicular, tubular, or septate appearance depending on the specific cell, state of cellular metabolism, and state of development.

    D. The inner membrane and **cristae** contain **globular particles** that have enzymes involved in **oxidative phosphorylation,** generation of **adenosine triphosphate (ATP).** Proteins of the **respiratory electron transfer chain** are located on the inner mitochondrial membrane.

    E. The **matrix** is the space between the cristae that contains **dense granules,** which may bind calcium ions.

MITOCHONDRIAL DISEASE AND INVOLVEMENT IN CELL DEATH

- *Leber's hereditary optic neuropathy* is a neurodegenerative disease, occurring usually in males, which results from a mutation of the mitochondrial gene involved in the generation of ATP.
- This disease is characterized by degeneration of the optic nerve, resulting in central vision loss.
- In mitochondria of normal cells, **cytochrome c** is found in the mitochondrial intermembrane space.
- After stimuli that trigger cell death, among them receptor ligand interactions, withdrawal of growth factors, and cytotoxic T cells, mitochondria become damaged, releasing **cytochrome c** into the cytoplasm.
- Cytochrome c forms a complex with **caspase-9** and **pro-apoptotic protease activating factor (Apaf-1)**, which activates caspase-9 promoting **apoptosis,** a form of **programmed cell death.**

X. Lysosomes

    A. **Primary lysosomes** are small membrane-bound vesicles, packaged in the Golgi apparatus, that contain hydrolytic enzymes that digest intracellular components (Figure 1–1E).

    B. **Secondary lysosomes,** or **phagolysosomes,** are formed by the fusion of a primary lysosome and a phagocytized foreign body.

        1. This process is called **heterophagy.**

        2. With aging, these lysosomal structures can accumulate in cells of organs such as the kidney and brain and are sometimes called **residual bodies.**

    C. **Primary lysosomes** are 25–50 μm in diameter, whereas **residual bodies** are several times as large.

    D. After cell injury, damaged organelles and other cytoplasmic material become packaged in **autophagic vacuoles.** These vacuoles fuse with primary lysosomes to form **autophagosomes** in a process called **autophagy.**

LYSOSOMAL STORAGE DISEASES

- The lysosomal storage diseases, the enzyme or enzymes that are deficient in each disease, and the associated clinical symptoms are shown in Table 1–2.

XI. Peroxisomes

    A. **Peroxisomes** are membrane-bound vesicles, spheroid or ovoid in shape and ~0.5 μm in diameter.

    B. These organelles contain 50 enzymes such as **catalase** and **oxidative enzymes,** including urate oxidase.

    C. **Catalase** breaks down hydrogen peroxide and synthesizes bile acids and plasmalogens that are found in the brain and heart.

Table 1–2. Lysosomal storage diseases.

Disease	Deficient Enzyme	Symptoms
Fabry's disease[a]	α-galactosidase A	Renal failure
Gaucher's disease[b]	β-glucocerebrosidase	Enlarged liver Pain in bone marrow
Hermansky-Pudlak syndrome[a]	Ceroid, pigment[c]	Albinism Cardiomyopathy Visual impairment
Hunter's syndrome[a]	Iduronate sulfatase	Mild mental retardation
Hurler's disease[b]	α-L-iduronidase	Mental retardation Corneal disorder
I-cell disease[b]	UDP-N-acetyl glucosamine N-acetylglucosaminyl-1-phosphotransferase	Graded I–IV Mental retardation Retinal degeneration
Krabbe's disease[b]	Galactosylceramide β-galactosidase	Optic atrophy
Niemann-Pick disease[b]	Sphingomyelinase	Death by 3 years
Tay-Sachs disease[b]	Hexosaminidase A	Swelling neurons of central nervous system and retina Cherry spot at macula

[a]X-linked recessive.
[b]Autosomal recessive.
[c]Accumulates in lysosomes.

PEROXISOME DISEASE

- **Zellweger syndrome,** also called **cerebrohepatorenal syndrome,** is caused by an absence of **peroxisomes,** primarily in the liver and kidney.
- This disease results in craniofacial abnormalities, polycystic kidneys, hepatomegaly, and infant death.

CLINICAL PROBLEMS

1. Which of the following components of mitochondria would be most abundant in those cells that have high ATP production?

A. Matrix

B. Cristae

C. Outer mitochondrial membrane

D. Dense granules

E. Inner membrane space

2. Which of the following events occurs during anaphase of mitosis?

   A. Chromatids separate

   B. Chromosomes coil

   C. Nuclear envelope reforms

   D. Chromosomes uncoil

   E. Nuclear envelope begins to break down

After examining a tissue from a biopsy specimen, a pathologist notes that a large percentage of the cells have nuclei with abundant nucleoli.

3. Which of the following cellular events would be elevated in these cells?

   A. Mitosis

   B. mRNA production

   C. Production of secreted proteins

   D. rRNA production

   E. Exocytosis

Your patient exhibits dwarfism and infantile sexual development. Karyotypic analysis reveals a single X chromosome but no other chromosomal anomalies.

4. This karyotype would be associated with which of the following genetic disorders?

   A. Klinefelter's syndrome

   B. Turner's syndrome

   C. Patau's syndrome

   D. Monogolism

   E. Edwards' syndrome

5. Cells that are actively involved in the phagocytosis of extracellular material would contain high levels of which of the following cellular organelles?

   A. Rough endoplasmic reticulum

   B. Golgi apparatus

   C. Lysosomes

   D. Smooth endoplasmic reticulum

   E. Ribosomes

6. Activation of which of the following would trigger a cell to enter the mitotic phase of the cell cycle?

A. Cyclin D-Cdc2

   B. Retinoblastoma gene

   C. p53 gene

   D. Cyclin G_1-Cdc2

   E. Cyclin B-Cdc2

An infant boy suffering from Zellweger syndrome exhibits craniofacial abnormalities and polycystic kidneys. He dies in early infancy. Biochemical assays showed that the spleen and liver have reduced levels of oxidases.

7. Which of the following organelles was responsible for this disorder?

   A. Lysosomes

   B. Peroxisomes

   C. Golgi apparatus

   D. Mitochondria

   E. Rough endoplasmic reticulum

A small tumor was excised from the adrenal gland of an adult male. After examination, the cells of the tumor were shown to express excessively high levels of steroid hormones.

8. Which of the following organelles were abundant within these tumor cells?

   A. Smooth endoplasmic reticulum

   B. Rough endoplasmic reticulum

   C. Golgi apparatus

   D. Lysosomes

   E. Peroxisomes

A tumor was removed from the liver of a patient, and cell cultures of tumor cells were established. Before chromosomal analysis of these cells, colchicine was added to a culture. This alkaloid prevents the polymerization of microtubules.

9. These cancerous cells would be arrested at which of the following stages of mitosis?

   A. Anaphase

   B. Interphase

   C. Metaphase

   D. Telophase

   E. Prophase

10. Which of the following structures are constructed mainly of the myofilament actin?

   A. Microtubules

   B. Basal bodies

   C. Intermediate filaments

   D. Cilia

   E. Neurofilaments

A fresh biopsy specimen appears to be undergoing atrophy. The nuclei of cells of this tissue are isolated by fractionation and homogenized. The resultant nuclear material is separated in an agarose gel and stained with ethidium bromide, which binds thymidine residues of DNA. Examining the gel under ultraviolet light, the researcher notes that the DNA is separated in increments of 200 base pairs by comparing it with the DNA standards.

11. Which of the following accounts for this observation?

    A. Chromosomes

    B. Heterochromatin

    C. Nucleoli

    D. Euchromatin

    E. Nucleosomes

A 50-year-old male develops antibodies to lamins as shown by Western blot analysis of his blood serum. The patient develops clinical conditions that are not affected by known medications. You want to determine what effect the antibody has on cells.

12. Which of the following cellular functions would be adversely affected by the nonfunctional lamins?

    A. Exocytosis

    B. Formation of nuclear envelope

    C. Phagocytosis

    D. Fusion of Golgi vesicles with cell membrane

    E. Chromatin condensation

A pathologist is studying a tissue section of the cerebrum by light microscopy. He notes a structure within a neuron that has both a granular and a fibrous component.

13. Which of the following is this pathologist examining?

    A. Ribosome

    B. Golgi apparatus

    C. Nucleolus

    D. Rough endoplasmic reticulum

    E. Mitochondria

14. Based on question 13, what is the main function of the cellular component?

    A. ATP production

    B. Protein trafficking

    C. Protein packaging

    D. rRNA production

    E. Site of protein synthesis

ANSWERS

1. The answer is B. Cristae (and the inner mitochondrial membrane) of mitochondria contain globular structures that have ATP synthase activity, which generates ATP from adenosine diphosphate (ADP).

2. The answer is A. Anaphase is marked by the separation of sister chromatids after the chromosomes are arranged at the equatorial plate. After separation, the chromatids migrate to opposite poles of the cell.

3. The answer is D. Nucleoli function in the production of rRNA, which forms the small and large subunits of ribosomes.

4. The answer is B. Loss of an X chromosome results in Turner's syndrome. The other genetic disorders described are characterized by an increased number of chromosomes.

5. The answer is C. Cells such as macrophages and neutrophils phagocytize extracellular or intracellular material in tissues or blood, respectively. In order for these cells to degrade the internalized material, lysosomes containing enzymes are required.

6. The answer is E. To enter the mitosis, an M phase-promoting factor must be formed, which is a complex consisting of cyclin B-Cdc2.

7. The answer is B. Zellweger syndrome is marked by a decrease in the number of peroxisomes.

8. The answer is A. Smooth endoplasmic reticulum (SER) has several functions, one of which is the synthesis of steroid hormones. SER is abundant in all steroid-producing cells such as within the liver, testis, and adrenal cortex.

9. The answer is C. Failure of microtubules to form or polymerize prevents chromatids of each chromosome from separating in anaphase. Thus, these cells would be arrested in metaphase of the mitosis.

10. The answer is D. Cilia have a central core of actin myofilaments.

11. The answer is E. Nucleosomes consist of histones and DNA in 200 base pair fragments. Thus, the researcher was noting DNA that was cleaved at the DNA linker site between adjacent nucleosomes. This observation is typical of cell undergoing apoptosis.

12. The answer is B. Lamins play a critical role in the reformation of the nuclear envelope during the later stages of telophase of mitosis. Their inability to polymerize or reform would adversely affect nuclear envelope formation.

13. The answer is C. The nucleolus consists of 2 regions: the pars granulosa and pars fibrosa.

14. The answer is D. The nucleolus functions as the site of rRNA synthesis.

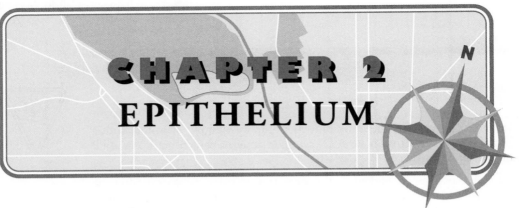

CHAPTER 2
EPITHELIUM

I. Four Primary Tissues

    A. Tissues are a group of similar cells specialized to perform a common function.

    B. The **4 basic tissues** in the human body are **epithelial, connective, muscle,** and **nervous.**

    C. All the **organs** are composed, to different degrees, of these 4 basic tissues with the one exception that the central nervous system does not contain connective tissue.

II. Epithelial Tissue

    A. Epithelial tissue consists of closely apposed cells interconnected by **junctional complexes,** with very little intercellular substance.

    B. Epithelia are **avascular** and thus are dependent on the blood vessels of the underlying connective tissue for nutrients.

    C. This tissue is arranged as sheets or membranes to either cover or line surfaces of the body or to form **glands.**

    D. All molecules or substances that enter or exit the body must pass through an epithelial layer.

        1. In many cases epithelial layers can modify and change this passage, thus influencing normal homeostatic mechanisms.

        2. Some epithelial membranes protect the body, whereas others allow absorption to occur efficiently.

    E. Epithelial tissue has the ability to shed and renew.

        1. This **turnover rate** varies from a few days in the intestine to months and years as in the ductus deferens of the male.

        2. The **mitotically active** cells lie in the basal region of epithelia, adjacent to its basal lamina.

III. Classification of Epithelial Tissue

    A. Epithelial tissues are classified according to both the **shape** of cells and the **number** of cell layers (Figure 2–1).

    B. Classifications based on the **shape of epithelial cells** are **squamous** (flat) cells, **cuboidal** (cube shaped) cells, and **columnar** (pillar shaped) cells.

    C. Classification based on the **number of cell layers** includes simple and stratified layers.

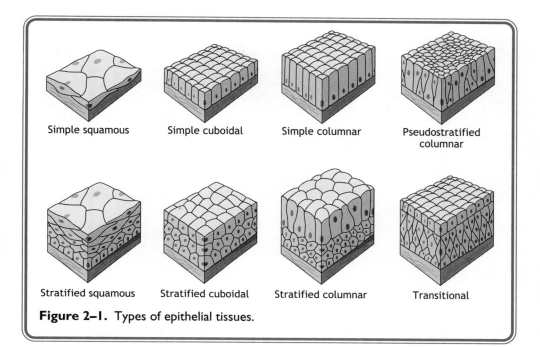

Figure 2–1. Types of epithelial tissues.

Labels in figure (top row): Simple squamous, Simple cuboidal, Simple columnar, Pseudostratified columnar

Labels in figure (bottom row): Stratified squamous, Stratified cuboidal, Stratified columnar, Transitional

     1. A simple epithelium consists of a single layer of cells, which rest on a basal lamina and may be **simple squamous, simple cuboidal,** or **simple columnar.**

     2. A stratified epithelium consists of 2 or more layers of cells, all of which rest on an underlying basal lamina and may be **stratified squamous, stratified cuboidal,** or **stratified columnar.**

  D. Specialized epithelial layers include pseudostratified columnar and transitional.

     1. Pseudostratified columnar epithelium is a single layer of columnar epithelial cells that only appears stratified.

        a. This apparent stratification results from all the cells residing on the basal lamina, but not all of them reach the surface (lumen).

        b. This type of epithelium is frequently ciliated and is found lining the lumen of the **trachea.**

     2. Transitional epithelium is a stratified epithelial layer found lining the renal major calyx and **urinary tract.**

        a. The apical cells are not classified as squamous, cuboidal, or columnar but appear round or oval.

        b. Some of these apical cells are **binucleate.**

IV. Function and Distribution of Epithelia

  A. The function of **simple squamous epithelium** is threefold.

     1. They allow gaseous exchange as in the **alveoli** of the lung.

     2. They act as a lining membrane, such as the **parietal layer of Bowman's capsule** in the kidney, which is called **endothelium** when lining blood vessels and **mesothelium** when lining the ventral surface of organs (eg, liver and kidney).

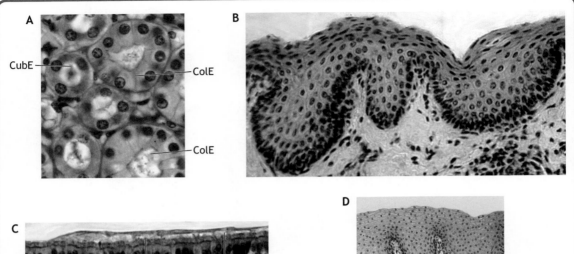

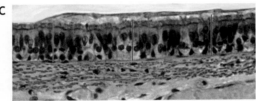

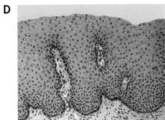

Figure 2–2. Photomicrographs of epithelial tissues. **A:** Simple cuboidal (CubE) and columnar epithelium (ColE). **B:** Stratified, nonkeratinized squamous epithelium. **C:** Pseudostratified ciliated columnar epithelium. **D:** Transitional epithelium.

3. They act as a lubricating membrane such as the **visceral layer of the lung** and the **pericardium** of the heart.

B. The function of the **simple cuboidal epithelium** is twofold (Figure 2–2A).
 1. It aids in absorption (eg, in the **distal convoluted tubules** in the renal cortex).
 2. It aids in secretion (eg, in the **secretory segment of salivary glands**).

C. The function of the **simple columnar epithelium** is also twofold (Figure 2–2A).
 1. It aids in absorption, such as the lumen of the **gallbladder.**
 2. It aids in transportation and absorption with cilia (eg, lining the lumen of the **intestine**).

D. The function of the **stratified squamous epithelium** is twofold.
 1. It aids in protection and secretion in its nonkeratinized form, lining the lumen of the **vagina** and **esophagus** (Figure 2–2B).
 2. It aids in protection in its keratinized form, in the **epidermis** of skin.

METAPLASIA

- *The transformation of one tissue type to another is called **metaplasia.***
- *A stratified squamous epithelium may replace other epithelial types in areas of chronic irritation.*

- In individuals who smoke, the normal simple columnar epithelium lining the bronchus is replaced by a nonkeratinized, stratified squamous epithelium, demonstrating the great lengths to which the body will go to adapt and protect itself.

BARRETT'S ESOPHAGUS

- A change in the epithelial lining of the esophagus from a stratified squamous, nonkeratinized epithelium to a mucus-secreting columnar epithelium is a form of **metaplasia.**
- This condition, called **Barrett's esophagus,** is caused by chronic acid reflux.

MALIGNANT AND BENIGN GROWTHS

- A malignant neoplasm, an abnormal or uncontrolled growth, made up of epithelial cells, is called a **carcinoma.** Of all epithelial tissues, stratified squamous is the most likely to undergo malignant changes, producing a **squamous cell carcinoma.**
- **Basal cell carcinoma** is a slow-growing tumor of basal cells of the epidermis that do not usually metastasize.
- Benign epithelial growths include protrusions called **polyps,** finger-like projections called **papillomas,** and cyst-appearing tumors known as **cystadenomas.**
- **Seborrheic keratoses** are round, plaque-appearing epithelial tumors that are most frequently found on the extremities, head, and neck.

    E. The **stratified cuboidal epithelium** functions in secretion and absorption (eg, in the lining of ducts of **sweat glands**).

    F. The function of the **stratified columnar epithelium** is twofold.
      1. It aids in protection, such as the **conjunctiva** covering the eyeball and male urethra.
      2. It aids in secretion (eg, in the **ducts** of large glands).

    G. The **pseudostratified ciliated columnar epithelium** functions in transportation, secretion, and lubrication in the linings of the lumen of the **trachea** and main and secondary **bronchi** (Figure 2–2C).

    H. **Transitional epithelium** functions in stretching and protection in the linings of the **urinary bladder, ureter,** and **major calyces** (Figure 2–2D).

HYPERPLASIA AND HYPOPLASIA

- **Hyperplasia** is a condition in which the number of cells increases in an organ or tissue with a possible increase in its bulk, whereas **hypoplasia** is a decrease in cell number.

HYPERTROPHY

- **Hypertrophy** is a condition of an increase in tissue size without increase in cell number.

ATROPHY

- **Atrophy** is a condition in which a tissue undergoes death with a decreased cellular volume.

  V. Epithelial Glands
    A. Exocrine Glands
      1. These glands develop as **down-growths** of an epithelial membrane that secrete onto the surface of the epithelium via ducts (Figure 2–3).

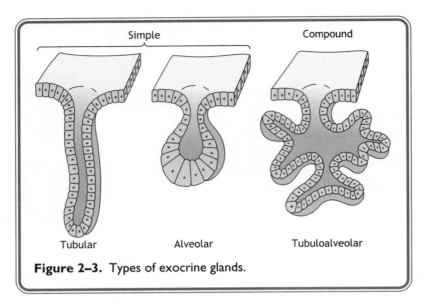

Figure 2–3. Types of exocrine glands.

    2. Exocrine glands are **simple or compound, branched or unbranched,** and **tubular or alveolar** in shape.
        a. These glands are found in the digestive, respiratory, urinary, and reproductive systems.
        b. The **goblet cell** is a unicellular exocrine gland that is found in the luminal epithelium of the digestive and respiratory tracts.

  B. Endocrine Glands
    1. Endocrine glands develop as exocrine glands (Figure 2–3) but **lose their ductal connection** to the epithelial surface.
    2. These glands release hormones into the surrounding connective tissue where these molecules enter blood vessels.
    3. Examples of endocrine glands are **adrenal, pituitary, thyroid, and pineal glands** as well as clusters of cells within the **pancreas (islets of Langerhans)** and **testes (interstitial cells of Leydig).**

ADENOMA AND ADENOCARCINOMA

CLINICAL CORRELATION

- *The glands of the body may also undergo **neoplastic** changes. A benign epithelial growth with glandular morphology is called an **adenoma,** whereas a malignant growth of this type is called **adenocarcinoma** (eg, **cervical adenocarcinoma**).*

VI. Cell Surface Specializations

  A. Cilia and Flagella
    1. Motile cell apical projections that originate from **basal bodies** in the cytoplasm are called **cilia.**
    2. A cilium is formed of a central axoneme that consists of a 9 + 2 arrangement of **microtubules.**

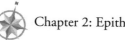

3. A **microtubule-associated protein,** with **adenosine triphosphatase (AT-Pase) activity,** called **dynein,** extends from one member of the doublet to another, and ciliary movement is dependent on the activity of this protein.
4. These structures function in the movement of ova in the **uterine tube** and mucus in the **upper respiratory tract.**
5. A **flagellum** is similar in structure to a cilium but is longer and is limited to 1 per cell: the **spermatozoa.**

KARTAGENER'S SYNDROME

- *An abnormal or absent ciliary beat resulting from a lack of **dynein** cross-arms is called **Kartagener's syndrome** (immotile cilia syndrome). This condition results in infections of the respiratory tract, possible infertility in females, and sterility in males.*

   B. Stereocilia

      1. **Stereocilia** are long cilia that contain a central core of **actin** rather than axonemes and thus are not true cilia.
      2. Stereocilia project from the apical surface of inner and outer hair cells of the **organ of Corti,** where they are involved in the generation of action potentials by sound waves.
      3. These structures project from cells lining the **epididymis** and **ductus deferens** in the male where they function in **absorption** and assisting the **maturation of sperm.**

   C. Microvilli
      1. The free surface of the outer epithelial cells may exhibit small, immobile cytoplasm projections called **microvilli.**
      2. These specializations increase the surface area and thus the **absorptive capacity** of epithelial cells, especially those lining the intestinal tract.

CELIAC SPRUE

- ***Celiac sprue,*** *or nontropical sprue, results from sensitivity to **gluten,** which is a component of wheat flour.*
- *This disease is characterized by a **loss of microvilli** on surface epithelium in the small intestine. This microvilli abnormality leads to a reduced absorptive capacity with resultant **osmotic diarrhea.***

   D. Basal Lamina
      1. A **basal lamina** underlies the basal surface of epithelial cells that **separates** epithelia from the supportive connective tissue.
      2. This layer is composed of a **lamina densa** and **lamina rara** (lamina lucida).
      3. The basal lamina **regulates** the exchange of macromolecules between the epithelium and the underlying connective tissue, such as between the epidermis of skin and the underlying dermis.
      4. **Collagen type IV,** laminin, fibronectin, and heparin sulfate are the major components of a basal lamina.
      5. A **basement membrane** is formed by the fusion of 2 basal laminae from 2 adjacent tissues.

VII. Cell–Cell Adhesion

   A. Zonulae Occludens (Tight Junctions)
      1. **Zonulae occludens** (Figure 2–4) are intercellular junctions that maintain **cell adhesion** and communication between adjacent epithelial cells.

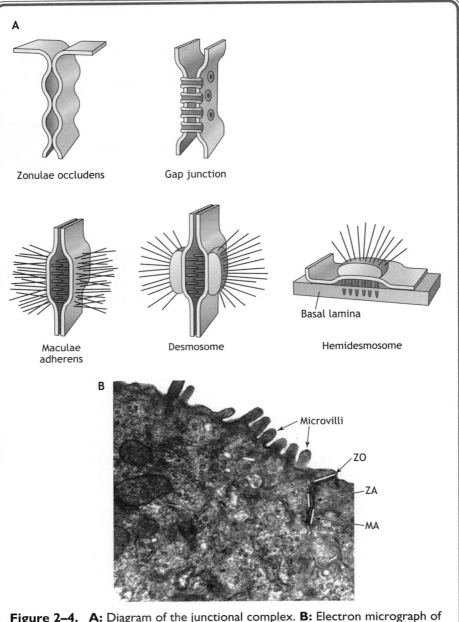

Figure 2–4. A: Diagram of the junctional complex. **B:** Electron micrograph of the junctional complex. (ZO = zonulae occludens; ZA = zonulae adherens; MA = maculae adherens.)

2. **Occludin** and **claudins** are transmembrane proteins that form a belt around adjacent cells.

3. Contact between occludin and claudins is reinforced by **cadherins** and 3 **zonulae occludens** proteins.

4. These junctions prevent molecules from passing between adjacent cells, especially in the gastrointestinal tract, and are essential in forming the **blood–brain, cerebrospinal fluid–brain,** and **blood–testis barriers.**

B. **Zonulae Adherens**

1. Zonulae adherens are intercellular junctions that encircle cells found immediately basal to the tight junction.

2. The connection between cells is maintained extracellularly by **cadherins** and intracellularly by **vinculin, catenins, α-actinin, plakoglobulin,** and actin filaments.

C. **Maculae Adherens**

1. These structures are also called **desmosomes** or **spot welds** because **plaques** are distributed along the lateral surface of adjacent epithelial cells. These junctions maintain cell adhesion between cells of the **epidermis** of skin, especially demonstrated in the stratum spinosum.

2. The **plaques,** consisting of **plakoglobins** and **desmoplakins,** are anchored by cytokeratins, which is a class of **intermediate filament proteins.**

D. **Hemidesmosome**

1. **Hemidesmosomes** are adhering junctions consisting of **cytokeratin tonofilaments** attached to a basal membrane surface of a cell found in the basal layer of the epidermis (**stratum basale**).

2. These structures anchor cells to components of a basement membrane by interacting with **integrin** transmembrane proteins.

BULLOUS PEMPHIGOID

- *Antibodies produced against transmembrane linker glycoproteins of **hemidesmosomes** cause **blisters**, which occur at the junction of the epidermis and dermis. This condition is called **bullous pemphigoid**.*

E. **Gap Junctions**

1. Gap junctions consist of 6 very closely packed **connexin** transmembrane proteins.

2. These complexes produce a **channel,** which allows passage of amino acids, ions, and proteins of less than **1 kDa** in molecular weight between adjacent cells.

3. Gap junctions are particularly important in **cardiac muscle,** allowing movement of **calcium ions.**

MALIGNANT CELL GROWTH

- *Cancerous cells typically lack **gap junctions**; thus, these cells are unable to communicate or interact intercellularly, which would regulate cellular activities. This gap junction defect can result in **uncontrolled mitotic activity** and **tumor growth**.*

CLINICAL PROBLEMS

A 40-year-old man has a 50-pack/year history of smoking. He complains of mucus accumulation and constant cough. A biopsy reveals that the epithelial lining of the tracheal lumen is a stratified squamous epithelium.

1. Which of the following terms describes the process of the change to another epithelial type?

   A. Metastasis

   B. Neoplasia

   C. Metaplasia

   D. Hyperplasia

   E. Hypoplasia

2. Based on question 1, what type of epithelium normally lines the lumen of the trachea?

   A. Stratified columnar epithelium

   B. Pseudostratified ciliated columnar epithelium

   C. Transitional epithelium

   D. Simple squamous epithelium

   E. Simple columnar epithelium

A patient complains of pain in the abdomen. It is determined that the pain was a result of large substances passing between cells that line the intestinal lumen directly into the underlying connective tissue.

3. Which of the following types of junctions were absent or not completely functional?

   A. Zonulae occludens

   B. Maculae adherens

   C. Zonulae adherens

   D. Gap junctions

   E. Hemidesmosome

4. Junctions are essential in maintaining the close association of the cells within the epidermis. Which of the following junctions is important in maintaining cell adherence in all layers of the epidermis?

   A. Gap junctions

   B. Hemidesmosome

   C. Zonulae adherens

   D. Zonulae occludens

   E. Maculae adherens

5. Which of the following proteins aids in maintaining the close association of the cells of the epidermis of the skin?

   A. Cadherins

   B. Occludins

   C. Intermediate filaments

   D. Vinculin

   E. Collagen

6. Simple squamous epithelium is prevalent throughout the body and organ systems. Which of the following is a primary function of this type of epithelium?

   A. Secretion

   B. Lubrication

   C. Absorption

   D. Protection

   E. Excretion

A biopsy of the urinary bladder is done in a patient who has complained of lower abdominal pain. In the pathologist's report, the luminal epithelium is described as normal.

7. What type of epithelium did the pathologist observe?

   A. Stratified squamous epithelium

   B. Simple squamous nonkeratinized epithelium

   C. Pseudostratified ciliated columnar epithelium

   D. Transitional epithelium

   E. Simple columnar epithelium

8. Microvilli are essential components of epithelial cells of the small intestine. Which of the following functions would be defective resulting from a lack of microvilli on epithelia?

   A. Stretching

   B. Movement

   C. Protection

   D. Secretion

   E. Absorption

The layer of cells that lines the outer surface of the lungs can be irritated by inhaled asbestos particles. These cells can become cancerous and lethal.

9. Which of the following is the cell layer being described?

   A. Simple squamous mesothelium

   B. Simple squamous endothelium

   C. Transitional epithelium

   D. Simple cuboidal epithelium

   E. Simple columnar epithelium

10. Which of the following is a characteristic typical of simple layers of epithelia?
    A. All the cells border an open lumen
    B. All the cells function to prevent abrasion
    C. All the cells rest on a basal lamina
    D. All the cells are joined by zonula occludens
    E. All the cells have microvilli

ANSWERS

1. The answer is C. Metaplasia is an abnormal transformation of a differentiated cell into a second cell type.

2. The answer is B. The trachea is lined by a pseudostratified ciliated columnar epithelium with goblet cells.

3. The answer is A. The only intercellular junction that would totally prevent passage of substances between cells would be zonulae occludens. Maculae adherens and zonulae adherens function as cell–cell attachment, whereas hemidesmosomes attach cells to a basal lamina. Gap junctions allow communication between cells.

4. The answer is E. Maculae adherens interconnect all cells of the epidermis, even the most superficial cells. Hemidesmosomes only connect the cells to the basal lamina.

5. The answer is C. Intermediate filaments are inserted into the attachment plaque of desmosomes. This association creates a strong adhesion between cells.

6. The answer is B. An important function of simple squamous epithelium is in lubrication such as the pleura of the lungs and the pericardium of the heart.

7. The answer is D. The lumen of the urinary bladder, as well as that of the ureter and calyces of the kidney, is lined by a transitional epithelium. This tissue can undergo distension with urine accumulation within the bladder.

8. The answer is E. Microvilli function to increase the surface area of cells. In the intestine, this is important because increased surface area results in an increased absorptive capacity of the cells lining the lumen of this organ.

9. The answer is A. The mesothelium, a type of simple squamous epithelium, lines the surface of organs such as the lung. An endothelium lines blood vessels.

10. The answer is C. All epithelial cells of the simple types rest on a basal lamina. The mesothelium is a simple epithelium that lines organs but does not line a lumen. Not all epithelial cells have microvilli or are joined by zonula occludens. Some epithelia, but not all, function to limit abrasion.

CHAPTER 3
CONNECTIVE TISSUE

I. General Features

A. Connective tissue, which is derived from **mesenchyme,** provides **structural support** and **protection** for internal organs of the body.

B. Connective tissue is composed of living **cells** and a nonliving material called the **extracellular matrix,** which consists of an amorphous ground substance, protein fibers, and tissue fluid.

II. Extracellular Matrix

A. **Amorphous Intercellular Ground Substance**
   1. **Glycosaminoglycans** (GAGs) are polysaccharides of disaccharides.
      a. **Dermatan sulfate** is found in dermis, tendons, ligaments, and fibrous cartilage.
      b. **Chondroitin sulfate** is found in hyaline and elastic cartilage.
      c. **Keratin sulfate** is found in the cornea.
      d. **Heparan sulfate** is associated primarily with reticular fibers and basal laminae.
   2. With the exception of hyaluronic acid, GAGs are bound covalently to a protein core forming a **proteoglycan molecule.**
   3. **Proteoglycans** are intensely hydrated structures that act as a selective physical barrier to regulate nutrients, inhibit microorganisms, and store growth factors.
   4. Structural glycoproteins include **fibronectin, laminin,** and **chondronectin.**

HURLER'S SYNDROME

- ***Hurler's syndrome*** *is recognized by an abnormal accumulation of extracellular proteoglycans and secretion of **dermatan sulfate** and **heparan sulfate** into the urine.*
- *This disease leads to bone and cartilage abnormalities and mental retardation.*
- *This syndrome is associated with a deficiency of α-L-iduronidase, an enzyme that hydrolyzes residues of dermatan and heparan sulfate.*

B. **Fibers**
   1. **Collagen** is the most abundant protein in the human body and consists of more than 10 types (Table 3–1).
      a. **Procollagen** is enzymatically cleaved at the nonhelical domain to form **tropocollagen.**

Table 3–1. Primary types of collagen.

Collagen	Characteristic	Distribution
Type I	Most abundant collagen Widespread distribution Forms fibers and bundles	Bone, tendons, skin
Type II	Forms fibrils only	Hyaline and elastic cartilage
Type III	First collagen synthesized	Reticular fibers during wound healing
Type IV	Does not form fibers	Only in basal lamina
Type V	Does not form fibrils	Fetal amnion and chorion and muscle and tendons

      b. **Fibrils** of collagen show cross-striations resulting from overlapping of tropocollagen molecules.
      c. Cross-linkage of collagen fibrils by proteoglycans and fibril-associated collagens with interrupted triple helices (**FACIT**) collagen form **collagen fibers.**
  2. Reticular fibers are synthesized by fibroblasts and stain with silver stains.
      a. Reticular fibers are composed primarily of **collagen type III.**
      b. Thin reticular fibers form a network or framework such as in **hematopoietic organs,** including the spleen, lymph nodes, and red bone marrow.
  3. Elastic fibers consist of microfibril-associated glycoproteins and **fibrillin.**
      a. Elastic fibers exhibit extreme elasticity.
      b. This type of fiber is synthesized by **fibroblasts,** even though it does not contain collagen.

EHLERS-DANLOS SYNDROME

- *Ehlers-Danlos syndrome is a group of inherited disorders of defects in the synthesis and structure of collagen.*
- *Symptoms of this disorder include laxity of tendons and ligaments, bruising, and fragility of tissues.*

MARFAN SYNDROME

- *Marfan syndrome is a genetic disorder in which connective tissue is excessively elastic.*
- *This syndrome results from a mutation in the fibrillin gene.*
- *Anomalies characteristic of this disorder include aortic aneurysms, scoliosis, and ocular defects.*

III. Cells of Connective Tissue

  A. Cells of connective tissue produce **matrix,** and **fibers** of the matrix provide an immunological line of defense for the body against invasive organisms.
  B. Mesenchymal cells are the progenitor cells that give rise to cells of connective tissue, including **fibroblasts** (Figure 3–1A).

C. The majority of cells in connective tissue are **fibroblasts.**
  1. Fibroblasts are large, stellate cells that have a pale-staining cytoplasm with a large, ovoid nucleus and 1–4 nucleoli.
  2. These cells function in synthesis of fibers and intercellular ground substance, tissue repair, and wound healing.

SCLERODERMA

• **Scleroderma** *is a genetic disorder in which fibroblasts undergo uncontrolled growth and collagen synthesis.*

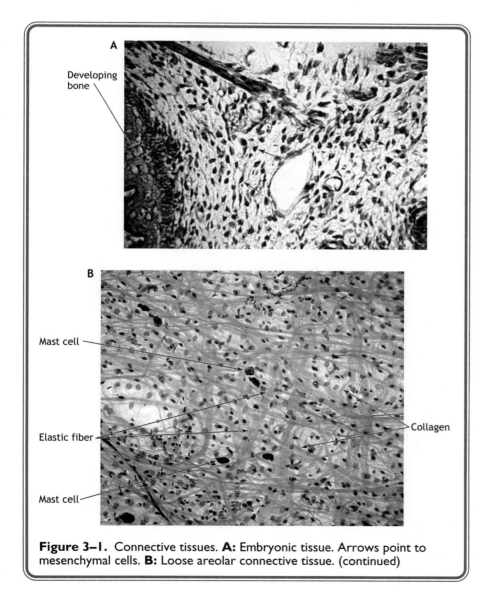

Figure 3–1. Connective tissues. **A:** Embryonic tissue. Arrows point to mesenchymal cells. **B:** Loose areolar connective tissue. (continued)

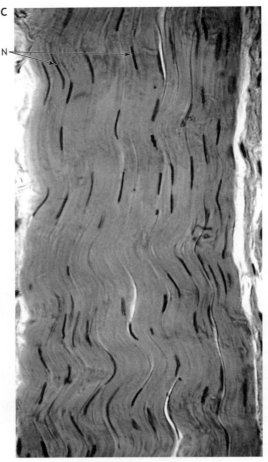

N

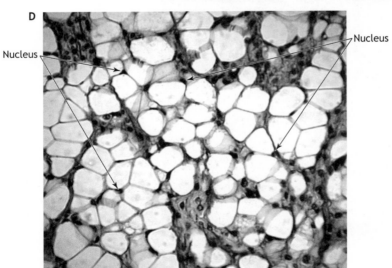

Nucleus

Nucleus

Figure 3–1. (continued) Connective tissues. **C:** Dense regular connective tissue. (N = nucleus). **D:** Adipose tissue.

- *This disorder results in **overproduction of connective tissue**, causing thick skin, impaired blood flow, and difficulty in swallowing.*

CONNECTIVE TISSUE INFLAMMATION

- *An **inflammatory reaction** of the body to foreign material is largely a connective tissue response involving these cells.*
- *The predominant connective tissue cell type, the **fibroblast**, is capable of synthesizing both the fibrous and amorphous matrix components and is essential for repair of the body after an injury.*

    D. Macrophages, also called **histiocytes,** function in defense through their phagocytic capacity.

        1. These cells are relatively large and have a cytoplasm that contains vacuoles and a spherical or indented nucleus, with 1–2 nucleoli.

        2. Monocytes differentiate into macrophages after migrating into connective tissue.

        3. Macrophages **secrete** several **factors,** including the interleukins, fibroblast growth factor, interferon, and colony-stimulating factors.

    E. Plasma cells are large, ovoid cells, with a densely basophilic cytoplasm and a clear area near the nucleus containing the **Golgi apparatus.**

        1. These cells, found in connective tissue, have a spherical and eccentrically placed nucleus with compact, **heterochromatin** alternating with light areas giving a **spoke-wheel appearance.**

        2. Plasma cells synthesize circulating **antibodies.**

        3. Plasma cells originate from **B lymphocytes** derived from bone marrow that migrate to nonthymic lymphoid tissues and then differentiate into plasma cells when activated.

    F. Mast cells are oval-shaped cells that have a cytoplasm containing basophilic granules and a small, spherical nucleus, most often central, with no apparent nucleoli (Figure 3–1B).

        1. Mast cells function as part of the **rapid immune response.**

        2. Mast cells have 2 secretion products: histamine and heparin.

            a. Histamine is a potent vasodilator that increases capillary permeability and causes smooth muscle contraction.

            b. Heparin is a blood anticoagulant, but its function within connective tissue is not known.

        3. Mast cells are found throughout the body but are most numerous in the **dermis** and within **loose connective tissue** of the digestive and respiratory tracts.

MAST CELL DISORDER

- *The granular **mast cell** appears to be important as an agent in many **immediate hypersensitivity reactions** such as hay fever.*
- *Hay fever is an allergic rhinitis marked by itching and lacrimation, sneezing, and irritation of nasal mucosa.*

    G. Adipose cells (adipocytes) are classified into 2 groups: white and brown.

        1. White (unilocular) adipose cells are extremely large, signet ring appearing and, as a group of cells, have a "chicken-wire" appearance.

            a. Nuclei of these cells are flattened and peripheral within a thin rim of cytoplasm containing a single lipid droplet.

b. White adipose cells originate from mesenchymally derived **lipoblasts,** which resemble fibroblasts but are able to accumulate fat in their cytoplasm.
   2. The cytoplasm of **brown adipose cells** contains multiple lipid droplets interspersed with mitochondria but little rough endoplasmic reticulum (RER).
   a. These cells originate from **lipoblasts.**
   b. The nucleus of brown adipose cells is eccentric but not peripheral as in white adipose cells.

IV. Classification of Connective Tissue

   A. Embryonic Connective Tissue
   1. Embryonic tissue (Figure 3–1A) has abundant amorphous ground substance composed chiefly of hyaluronic acid.
   2. It contains very few fibers, stellate-shaped mesenchymal cells, and is found adjacent to developing bone in the embryo.

   B. Connective Tissue Proper
   1. Loose (areolar) connective tissue (Figure 3–1B) is flexible and highly vascularized and has numerous fibroblasts and macrophages, with a few mast cells.
   a. This tissue contains collagen, elastic, and some reticular fibers.
   b. This tissue is found in **mesentery,** omentum, papillary layer of the **dermis, hypodermis,** and some glands.
   2. Dense irregular connective tissue has sparse ground substance and **few cells** and numerous collagen fibers irregularly arranged.
   a. This tissue is less flexible but far more resistant to stress than loose connective tissue.
   b. This tissue is found in **dermis** and **organ capsules.**
   3. Dense regular connective tissue (Figure 3–1C) has collagen bundles arranged in parallel; this tissue is found primarily in **tendons** and **ligaments.**

TUMORS OF CONNECTIVE TISSUE

CLINICAL CORRELATION

- *The **neoplasms** that arise from supportive connective tissue cells are called **sarcomas,** including fibrosarcoma and osteosarcoma.*
- *Tumors of cells in connective tissue are named for their cell of origin such as **plasmacytoma** and **lymphoma.***

   C. Elastic Tissue
   1. Elastic tissue is composed of bundles of thick, parallel elastic fibers.
   2. This tissue is present in the **ligamentum flavin** of the vertebral column.

   D. Reticular Tissue
   1. Reticular tissue provides the architectural framework of the myeloid and lymphoid **hematopoietic organs.**
   2. Reticular tissue is formed by a fine matrix of branched reticular fibers, consisting of **type III collagen,** secreted by fibroblasts.

   E. Adipose Tissue
   1. White adipose tissue (Figure 3–1D) serves as a reserve energy source through homeostatic storage and mobilization of triglycerides.
   a. This tissue also provides insulation and shock absorption.
   b. The yellow to white color of this adipose tissue depends on the concentration of **cartenoids** in the diet.

2. Cells of **brown adipose tissue** produce heat in the first months of postnatal life.
   a. This tissue accounts for 2% of the body weight at birth but is of limited extent in adults.
   b. The brown color in this tissue is due to large number of **mitochondria,** and this tissue has a rich blood supply.

CLINICAL PROBLEMS

1. Macrophages would be most abundant in which of the following tissues?
   A. Loose areolar connective tissue
   B. Dense regular connective tissue
   C. Brown adipose tissue
   D. Dense irregular connective tissue
   E. Embryonic tissue

A pathologist is examining a tissue section with an electron microscope. He notices a cell in connective tissue that has a prominent Golgi apparatus and a nucleus with heterochromatin arranged in a spoke-wheel fashion.

2. Which of the following cell types is being observed?
   A. Macrophage
   B. Plasma cell
   C. Fibroblast
   D. Mast cell
   E. Brown adipose cell

3. Based on question 2, which of the following is the function of the cell?
   A. Histamine synthesis
   B. Phagocytosis
   C. Lipid synthesis
   D. Antibody production
   E. Collagen synthesis

4. Which of the following provides a primary structural network in hematopoietic organs?
   A. Glycosaminoglycans
   B. Collagen fibers
   C. Elastic fibers
   D. Proteoglycans
   E. Reticular fibers

Your patient suffered a complete tear of biceps brachii muscle at its insertion site at the radial tuberosity. For proper treatment, you must identify the damaged tissue.

5. Which of the following tissues was damaged at the tear site?

    A. Reticular tissue

    B. Elastic tissue

    C. Loose areolar connective tissue

    D. Dense regular connective tissue

    E. Dense irregular connective tissue

6. Your patient suffers from an immediate hypersensitivity reaction. Which of the following cell types is responsible for this condition?

    A. Plasma cell

    B. Fibroblast

    C. Mast cell

    D. Macrophage

    E. Adipose cell

7. Which of the following types of collagen is the most widely distributed and abundant within the body?

    A. Type I collagen

    B. Type II collagen

    C. Type III collagen

    D. Type IV collagen

    E. Type V collagen

A 45-year-old man presented with increased mobility at major joints and elasticity of skin. Further examination revealed damage to cartilage and ligaments.

8. Based on this information, which of the following is the proper diagnosis?

    A. Hurler's syndrome

    B. Marfan syndrome

    C. Scleroderma

    D. Ehlers-Danlos syndrome

    E. Addison's disease

Immunocytochemistry is performed on a tissue section using an antibody specific for type IV collagen. The investigator notes staining within an organ that contains abundant and varied connective tissue.

9. Which of the following would be identified by the antibody to type IV collagen?

    A. Reticular fibers

    B. Basal lamina

    C. Elastic fibers

D. Loose areolar connective tissue

    E. Adipose cells

10. Which of the following microscopically best characterizes brown fat cells?

    A. Lack of mitochondria

    B. Peripheral flattened nucleus

    C. Multiple droplets of lipid

    D. Abundant RER

    E. Poor blood supply

ANSWERS

1. The answer is A. Tissue macrophages are found within loose areolar connective tissue, especially within the gastrointestinal tract.

2. The answer is B. The cell that is described is a plasma cell. This cell has a distinctive nucleus because of its heterochromatin arrangement and an abundant Golgi adjacent to the nucleus.

3. The answer is D. Plasma cells synthesize and secrete circulating antibodies.

4. The answer is E. Reticular fibers within hematopoietic organs provide a supportive role. In other tissues such as the lung, these fibers limit the amount of stretch that the lung tissue will undergo.

5. The answer is D. Muscle is attached to bone by a tendon, which is composed of dense regular connective tissue. Thus, this tissue would be damaged by detachment of the tendon of the biceps brachii muscle from its insertion site.

6. The answer is C. Immediate hypersensitivity reaction is marked by an immediate release of molecules such as histamine and leukotrienes from the mast cell. The mast cell undergoes a process of degranulation.

7. The answer is A. Type I collagen is the most abundant in the body found in skin, bone, and tendons.

8. The answer is D. Ehlers-Danlos syndrome results in a marked defect in the synthesis of connective tissue, specifically collagen. This disorder leads to anomalies such as skin and blood vessel fragility and joint problems.

9. The answer is B. Type IV collagen is exclusively found within the basal lamina, which serves as an anchor to epithelial cells. As other connective tissue components, it is synthesized and secreted by fibroblasts.

10. The answer is C. Brown fat cells have several lipid droplets and not a single droplet as white fat cells. The nucleus, although eccentric, is not flattened against the cell membrane. These cells have abundant mitochondria but little RER.

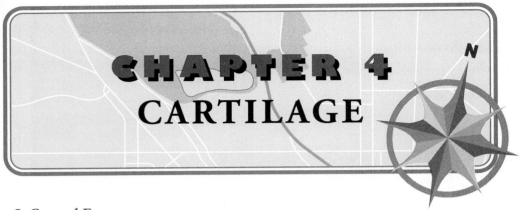

CHAPTER 4
CARTILAGE

I. General Features

A. **Cartilage** consists of fibers, cellular elements, and an amorphous ground substance that has protective and supportive roles.

B. In early embryonic development of cartilage, **appositional** and **interstitial cartilage** growth may occur together, whereas later growth is primarily appositional.

C. Cartilage is **avascular;** thus, nutrients, oxygen, and waste material are transported by passive diffusion through the matrix.

II. Cartilage Composition

A. Matrix and Amorphous Ground Substance

1. The **matrix** makes up **40%** of cartilage and primarily consists of **type II collagen** and **elastin.**

2. **Amorphous ground substances** constitute **60%** of cartilage, including glycosaminoglycans, chondroitin-6-sulfate and chondroitin-4-sulfate, hyaluronic acid, and keratin sulfate.

CAMPTOMELIC DYSPLASIA

• Type II collagen and the proteoglycan aggrecan are controlled by the transcription factor **Sox 9.** A mutation in this gene results in **camptomelic dysplasia,** which is marked by anomalies of the ribs and vertebral column and bowing of long bones.

VITAMIN DEFICIENCY DISORDERS

• Deficiency of **vitamin A** leads to anomalies at the epiphyseal plate, whereas excess vitamin A accelerates ossification at this site.

• **Scurvy** is caused by a vitamin C deficiency, which is marked by skeletal anomalies and hemorrhages. A major function of vitamin C is the activation of the hydroxylases of amino acids providing hydroxylation of **procollagen.**

B. Cells of Cartilage

1. Chondroblasts

a. **Chondroblasts** are found at the surface of cartilage, derived from **mesenchymal cells,** and are capable of elaborating a matrix.

b. These spindle-shaped cells have a small oval to round nucleus and a basophilic cytoplasm.

CHONDROMETAPLASIA

- *Metaplasia* of chondroblasts results in a condition called **chondrometaplasia.**
- *It occurs primarily in the synovial cartilage or tendinous sheaths.*

    2. Chondrocytes

        a. Chondrocytes are mature chondroblasts that have become completely enclosed in the territorial matrix.

        b. These cells, as **isogenous groups,** reside in a space within the matrix called a **lacuna.**

        c. Chondrocytes are round cells, with basophilic cytoplasm and a small round nucleus.

        d. Proliferating chondrocytes form the cartilaginous **epiphyseal plates** between the epiphysis and the shaft of a developing long bone.

ACHONDROPLASIA

- *Achondroplasia* is a genetic disorder in which the proliferation of chondrocytes is reduced at the epiphyseal plate. This condition results in **dwarfism** of the extremities and trunk.
- *Increased growth at the epiphyseal plate and its replacement by bone in children is controlled by the pituitary hormone* **somatotrophin.** *This hormone controls the synthesis of the liver hormone* **somatomedin C,** *which stimulates* **chondrocytes** *at the epiphyseal plate of long bones.*

    3. Chondroclasts are large cells responsible for absorption of cartilage.

  C. Perichondrium

    1. The outer layer of the **perichondrium,** which encloses hyaline and elastic cartilage, consists of collagen fibers and fibroblasts.

    2. The inner layer has fine collagen fibers, undifferentiated **mesenchymal cells,** and capillaries.

    3. Blood vessels and lymphatics of hyaline and elastic cartilage are supported within the fibers of the perichondrium.

  D. Growth of Cartilage

    1. Appositional growth occurs at the surface of the cartilage or perichondrium.

    2. Interstitial growth, which occurs between chondrocytes, is active during endochondral ossification.

    3. Mitosis is limited to 1 or 2 rounds of replication, which gives rise to **isogenous groups** of chondrocytes.

    4. Isogenous groups are surrounded by **territorial matrices,** which are separated by an **interterritorial matrix.**

III. Types of Cartilage

  A. Hyaline Cartilage

    1. Hyaline cartilage (Figure 4–1A) is found in the nasal passages, upper respiratory passage (including the larynx, trachea, and bronchi), costal cartilage of ribs, and articular cartilage at joints.

    2. This cartilage has a homogeneous staining matrix, which appears **bluish-white.**

    3. Chondrocytes are situated in **isogenous groups** in an amorphous matrix that contains fibrils of **type II collagen.**

    4. The matrix stains basophilic because of the high content of acidic sulfhydryl groups and carbohydrates.

Figure 4–1. Photomicrographs of cartilage. **A:** Hyaline cartilage. **B:** Elastic cartilage. (Pc = perichondrium; Ch = chondrocyte; IG = isogenous group; E = elastic fiber.)

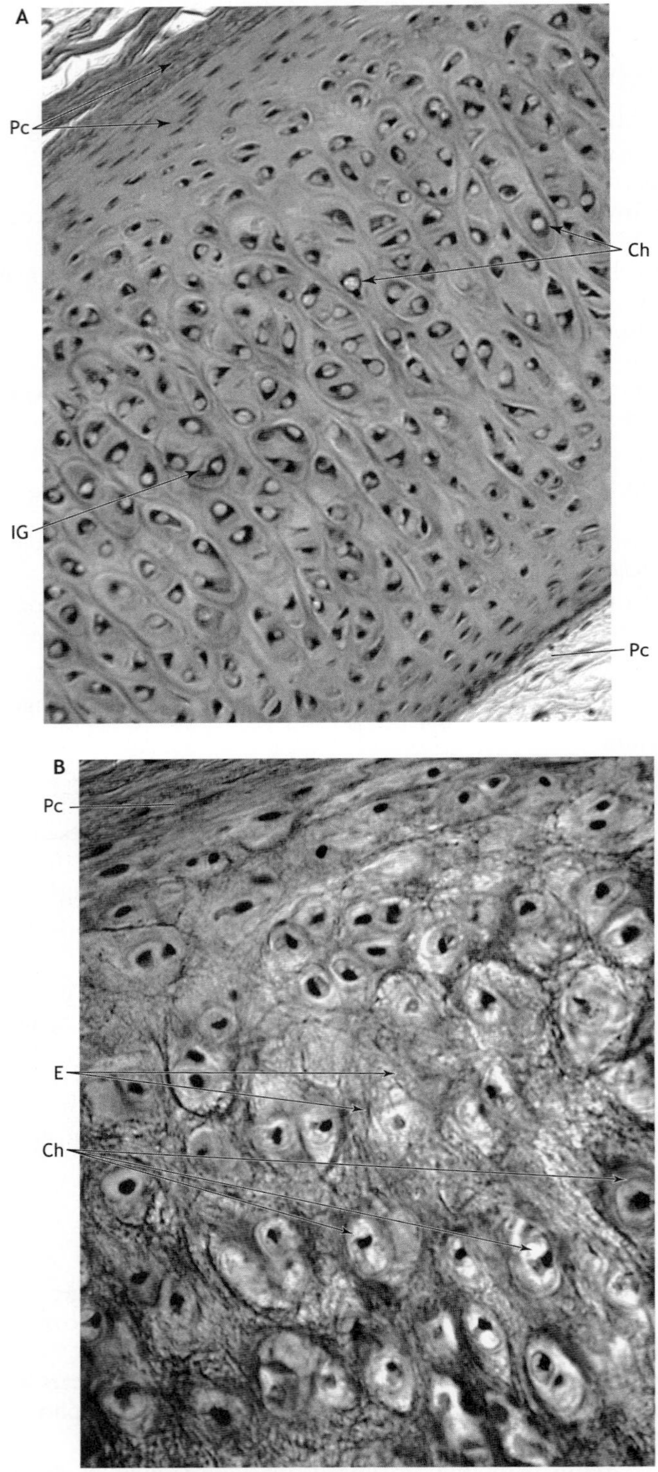

5. Hyaline cartilage has a low cell to matrix ratio, fewer cells than in other cartilage types, and the chondrocytes are evenly distributed.
6. Hyaline cartilage has a resiliency between elastic cartilage and fibrocartilage.

CHONDROMA

• A **chondroma** is a benign tumor of hyaline cartilage that develops within the substance of or at the periphery of cartilage.

CHONDROMALACIA

• **Chondromalacia** is a disease characterized by **softening** of articular cartilage, such as on the anterior surface of the patella and epiphyseal cartilage of stillborn fetuses.

B. Elastic Cartilage
1. **Elastic cartilage** (Figure 4–1B) is found in the pinna of ear, auditory (eustachian) tube, epiglottis, and corniculate and cuneiform cartilages of the larynx.
2. In this cartilage, the **elastic fibers** stain as a result of a high content of basic amino acids, and **type II collagen** is abundant.
3. Elastic cartilage is more opaque and the matrix less abundant than in hyaline cartilage.
4. Elastic cartilage has a high cell to matrix ratio and is the most resilient cartilage; thus, it affords a flexible form of support.

C. Fibrocartilage
1. **Fibrocartilage** is found in the intervertebral disks, cartilage at the pubic symphysis, and insertions of tendons and ligaments.
2. Chrondrocytes of fibrocartilage are often dispersed in a **linear arrangement,** not in isogenous groups
3. Fibrocartilage has a thin territorial matrix, whereas the interterritorial matrix consists of bundles of **type I collagen.**
4. This cartilage has the least resiliency of all cartilage but gives firm support and tensile strength.

CHONDROBLASTOMAS

• **Chondroblastomas** are benign tumors that are derived from immature cartilage cells. These tumors usually occur in the epiphyses of adolescents.

CHONDROSARCOMAS

• **Chondrosarcomas** are malignant tumors of cartilage cells or immediate precursor cells and occur in bones of the pelvis and shoulder of older individuals.

CLINICAL PROBLEMS

1. Which of the following provides the nutrients to mature chondrocytes?
   A. Vasculature of the cartilage
   B. Canals within the cartilage

C. Capillary network within the matrix

    D. Diffusion through the matrix

    E. Cellular interconnections between adjacent chondrocytes

2. Which of the following would best characterize fibrocartilage?

    A. Presence of a perichondrium

    B. Found at the epiphyseal plate

    C. Rows of chondrocytes

    D. Large, densely stained territorial matrix

    E. Lines bones at articular sites

3. Which of the following is the major component of the matrix of cartilage?

    A. Collagen

    B. Isogenous groups

    C. Glycosaminoglycans

    D. Chondrocytic processes

    E. Perichondrium

4. Chondroblasts are derived from which of the following cell types?

    A. Fibroblasts

    B. Proliferating chondrocytes

    C. Mesenchymal cells

    D. Cells within isogenous groups

    E. Cells of the perichondrium

5. Which of the following best characterizes hyaline cartilage?

    A. No perichondrium

    B. Single chondrocytes

    C. Found between vertebral bodies

    D. Least resilient of all forms of cartilage

    E. Abundant matrix

6. Elastic cartilage is found in which of the following body structures?

    A. Intervertebral disks

    B. Epiglottis

    C. Tracheal rings

    D. Cartilage at pubic symphysis

    E. Costal cartilage

7. Which of the following organelles plays an essential role in the sulfation of glycosaminoglycans within chondrocytes?

    A. Golgi apparatus

B. Polyribosomes

C. Smooth endoplasmic reticulum

D. Rough endoplasmic reticulum

E. Ribosomes associated with the nuclear envelope

8. Which of the following clinical conditions results from reduced mitotic activity of chondrocytes at the epiphyseal plate?

A. Chondroblastoma

B. Chondroma

C. Achondroplasia

D. Chondromalacia

E. Chondrometaplasia

ANSWERS

1. The answer is D. Cartilage is avascular; thus, nutrients supporting chondrocytes diffuse from blood vessels peripheral to this tissue.

2. The answer is C. Chondrocytes are arranged linearly, not in isogenous groups as in elastic and hyaline cartilage. Fibrocartilage does not have a perichondrium.

3. The answer is A. The matrix of cartilage primarily consists of collagen, whereas the ground substance consists of molecules such as glycosaminoglycans.

4. The answer is C. Chondroblasts are derived from mesenchymal cells. The perichondrium surrounding cartilage houses chondroblasts.

5. The answer is E. Hyaline cartilage has abundant matrix, with a perichondrium, and is more flexible than fibrocartilage.

6. The answer is B. Elastic cartilage is found in those structures or organs requiring flexibility, such as the epiglottis.

7. The answer is A. Sulfation of proteins occurs within the Golgi apparatus of chondrocytes.

8. The answer is C. Below-normal levels of chondrocytic mitotic activity result in the condition called achondroplasia.

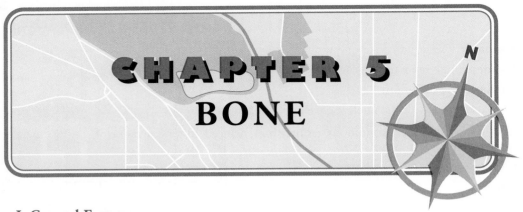

CHAPTER 5
BONE

I. General Features

A. Bone is a vascularized, rigid form of connective tissue.

B. Functions of bone include support, protection, reservoir for calcium ions, and locomotion in conjunction with skeletal muscle contraction.

II. Composition of Bone

A. Cell Types
   1. Osteoprogenitor cells
      a. **Osteoprogenitor cells** are mitotically active cells that are derived from **mesenchymal cells.**
      b. These immature cells are spindle shaped with a pale cytoplasm and are found adjacent to new-forming bone matrix.
      c. The nucleus of osteoprogenitor cells is pale staining.
   2. Osteoblasts
      a. **Osteoblasts** (Figure 5–1A) originate from osteoprogenitor cells and are squamous, cuboidal, or columnar in shape, with a basophilic cytoplasm.
      b. Osteoblasts synthesize fibers and amorphous ground substance, which is called **osteoid,** the mineralized component of bone.
      c. The proteins that control the differentiation of osteoblasts are **core-binding factors (Cbfa1), bone morphogenetic proteins (BMP),** and **transforming growth factor β** (TGF-β.).
   3. Osteocytes
      a. **Osteocytes** (Figure 5–1B) are the mature cells of bone, which reside in lacunae, surrounded by matrix, and function to maintain mineralized matrix.
      b. Osteocytes have a flattened oval shape and an oval nucleus.
      c. The cell processes that interconnect osteocytes are found within **canaliculi.**
      d. **Gap junctions** in osteocyte processes provide communication between these cells.
      e. **Osteocalcin,** a calcium-binding bone protein produced by osteocytes, is a serum marker for increased bone turnover in disease states.
   4. Osteoclasts
      a. **Osteoclasts** (Figure 5–1A) have a ruffled border and produce a surface depression in developing bone called **Howship's lacunae.**

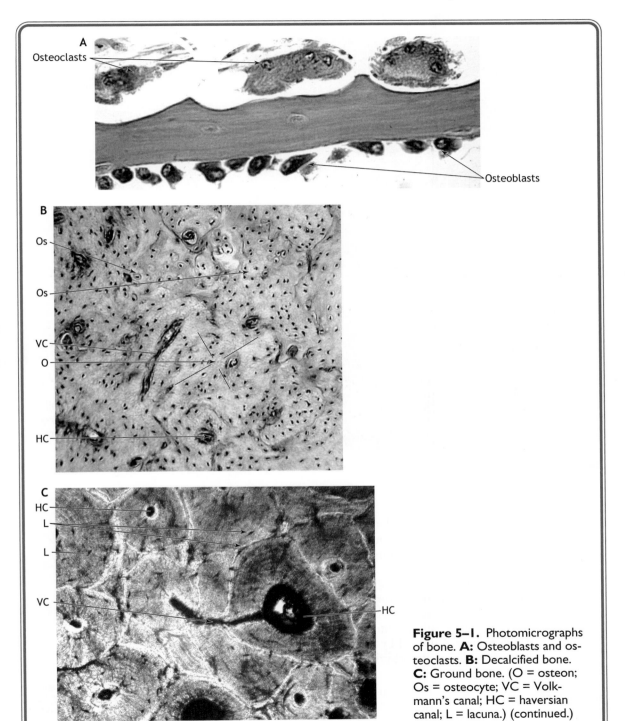

Figure 5–1. Photomicrographs of bone. **A:** Osteoblasts and osteoclasts. **B:** Decalcified bone. **C:** Ground bone. (O = osteon; Os = osteocyte; VC = Volkmann's canal; HC = haversian canal; L = lacuna.) (continued.)

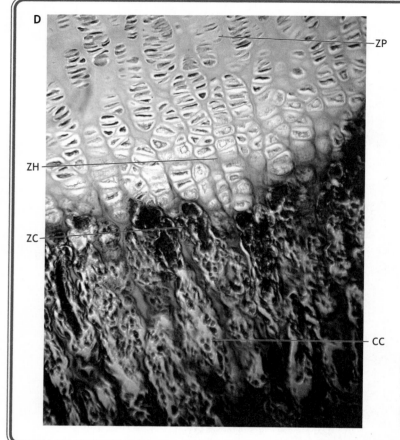

D

ZP

ZH

ZC

CC

Figure 5–1. Photomicrographs of bone (*continued*). **D:** Bone formation. (ZP = zone of proliferation; ZH = zone of hypertrophy; ZC = zone of calcification; CC = calcified cartilage.)

   b. These cells, usually found singly, are multinucleated because they originate from several **monocytes** and function in **resorption** of bone by release of H⁺ and secretion of lysosomal hydrolases, gelatinase, and collagenase.

   c. **Interleukin-1** stimulates osteoclast proliferation, whereas **osteoprotegerin,** produced by osteoblasts, inhibits osteoclast differentiation.

   d. **Parathyroid hormone** stimulates osteoblasts to secrete osteoclast-stimulating factor, which **activates** bone resorption by osteoclasts. This activity leads to the liberation of Ca^{2+} into the blood.

   e. **Calcitonin,** synthesized by parafollicular cells of the thyroid gland, acts on osteoclasts directly to **inhibit** bone resorption.

OSTEOPETROSIS

• **Osteopetrosis** is a genetic disorder characterized by an increased density of bone. This condition results from defective resorption of bone by dysfunctional osteoclasts.

CLINICAL CORRELATION

DWARFISM AND GIGANTISM

· **Somatotropin** *is a pituitary growth hormone that affects bone growth and remodeling. Deficiency of this hormone during childhood leads to **dwarfism,** whereas excess hormone before closure of the epiphyseal plate causes **gigantism.** Elevated hormone in adults causes thickening of bones such as of the jaw, fingers, and bones of the skull, called **acromegaly.***

   B. **Organic Components of Bone Matrix**
      1. Organic components are 35% of bone matrix and consist of **proteoglycans** such as chondroitin 4-sulfate, chondroitin 6-sulfate, and keratin 6-sulfate.
      2. **Type I collagen** fibers (90%) predominate in bone matrix.
      3. The glycoproteins such as **sialoprotein** and **osteocalcin** bind calcium and promote calcification.

   C. **Inorganic Components of Bone Matrix**
      1. Inorganic components comprise 65% of bone matrix and consist of calcium and phosphorus, which form crystals of **hydroxyapatite.**
      2. Other inorganic components of bone matrix include bicarbonate, citrate, magnesium, potassium, and sodium.

III. Types of Bone

   A. **Cancellous Bone**
      1. **Cancellous bone** consists of bony spicules or trabeculae that enclose bone marrow.
      2. This bone is found at the ends of long bones, called the **epiphysis,** which is separated from the diaphysis by the **epiphyseal plate.**

   B. **Compact Bone**
      1. **Compact bone** (Figures 5–1C, 5–2) appears as a solid mass in the **diaphysis** of the femur and other long bones.
      2. A **haversian system,** also called an **osteon,** is a concentrically arranged lamella of **compact bone** that surrounds a haversian canal.
      3. A **haversian canal** is a longitudinal space within the center of an osteon containing capillaries.
      4. **Volkmann's canal** is a channel that connects adjacent haversian canals. These canals contain blood vessels from the bone marrow and periosteum.
      5. The **outer circumferential lamellae** are positioned at the periphery of compact bone deep to the **periosteum.**
      6. The **inner circumferential lamellae** are on the internal surface adjacent to the **endosteum.**
      7. **Interstitial lamellae,** found between osteons, are separated by **cement lines.**
      8. The **periosteum** is a specialized connective tissue surrounding the periphery of bone that contains osteoprogenitor cells connected to compact bone by **Sharpey's fibers.**
      9. The **endosteum** is connective tissue that lines the **medullary canal** of bone containing **osteoprogenitor cells.**

IV. Endochondral Ossification

   A. **Endochondral ossification** (Figures 5–1D, 5–3) is bone formation using a cartilage model, which is gradually replaced by bone.

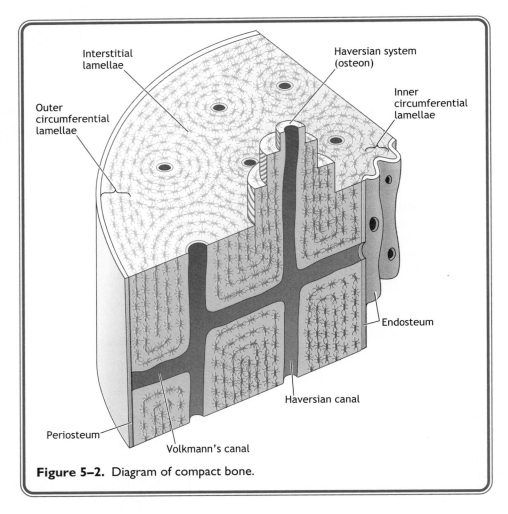

Interstitial lamellae

Haversian system (osteon)

Outer circumferential lamellae

Inner circumferential lamellae

Endosteum

Haversian canal

Periosteum

Volkmann's canal

Figure 5–2. Diagram of compact bone.

B. The first stage of endochondral ossification is the formation of **osteoblasts** around the cartilage.

C. In the second stage, a **primary center of ossification** is formed at the **diaphysis,** which is the site of increased bone **diameter.**

D. A **secondary center of ossification** is formed at the **epiphyses.**

E. The bony shaft, or diaphysis, is separated from the bony epiphyses by residual cartilages of the **epiphyseal plate,** which is the site of bone **longitudinal growth.**

F. There are 5 zones of osteogenesis.
   1. The **resting zone,** nearest to the epiphysis, is the site at which chondrocytes divide to form columns of differentiating cells.
   2. The **zone of proliferation** is the site of **mitotically** active chondrocytes. This activity lengthens the cartilage model.

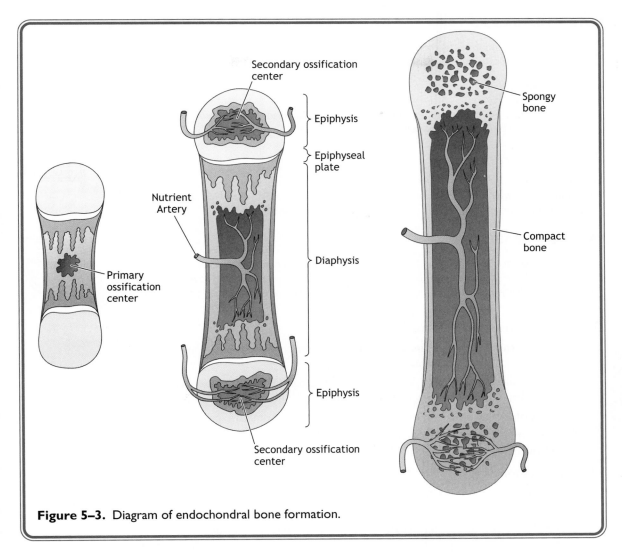

Figure 5–3. Diagram of endochondral bone formation.

3. The **zone of hypertrophy** has few mitotic cells, but the mature chondrocytes undergo hypertrophy, contain vacuoles, and accumulate **glycogen.**

4. The **zone of calcification** contains degenerating or dead chondrocytes. The matrix between these cells becomes filled with hydroxyapatite.

5. The **zone of ossification** contains differentiating osteoblasts that adhere to cartilaginous spicules and deposit thin layers of bone matrix.

RICKETS

• *Rickets, called **juvenile osteomalacia,** is an abnormal mineralization and development of the growth plate of bone leading to bone pain, growth retardation, and fatigability. This disorder results from **deficient vitamin D,** calcium, and phosphorus caused by malabsorption or dietary restrictions.*

OSTEOCHONDROSIS

- *Disease of the ossification centers of bone in children is called **osteochondrosis**. This disease begins as degeneration or necrosis that is followed by regeneration or recalcification.*

OSTEOPOROSIS

- *Loss of bone mass resulting in bone fragility and fractures is called **osteoporosis**. This condition is commonly caused by deficient **estrogen**.*

OSTEOMALACIA

- ***Osteomalacia** is the condition in which bone becomes progressively softened. This defect is due to a defect in mineralization of bone resulting from reduced **vitamin D**. This condition can result from reduced osteoblast activity while osteoclast activity is normal or even elevated.*

PAGET'S DISEASE

- ***Paget's disease** is characterized by excessive bone formation and breakdown, resulting in an increase in cancellous bone compared with compact bone. With this disease osteoblastic activity increases, whereas the activity of osteoclasts decreases. This disease usually affects the femur, pelvis, spine, and skull.*

V. Intramembranous Ossification

    A. During **intramembranous ossification, mesenchymal tissue** condenses into a primitive connective tissue.

    B. Stellate-shaped **mesenchymal cells** differentiate into **osteoprogenitor cells,** which become enclosed by its deposited bone.

    C. Osteoblasts become enclosed by bone matrix to form mature **osteocytes.**

    D. This type of bone is characteristic of the flat bones of the **skull, scapula,** and **manubrium.**

VI. Bone Remodeling

    A. During remodeling of compact bone, the matrix is gradually resorbed by **osteoclasts** and replaced by **osteoblasts.**

    B. The formation of new haversian systems usually takes 3–4 months.

    C. Turnover of cancellous bone by internal remodeling occurs at a rate of 26% per year, whereas the turnover rate of compact bone by remodeling is approximately 3% per year.

CLINICAL PROBLEMS

    1. Within compact bone, which of the following form connections between haversian systems?

       A. Howship's lacunae

       B. Volkmann's canals

C. Canaliculi

D. Lacunae

E. Cement lines

2. Cessation of growth at which of the following sites would result in no further longitudinal growth of long bones?

A. Endosteum

B. Primary center of ossification

C. Secondary center of ossification

D. Diaphysis

E. Epiphyseal plate

A blood analysis of your patient shows an increased level of parathyroid hormone.

3. Which of the following would result from increased levels of this hormone?

A. Increased mitotic activity of osteoprogenitor cells

B. Decreased serum calcium levels

C. Increased bone formation

D. Increased serum calcium levels

E. Increased hydroxyapatite crystal formation

4. Which of the following could be administered to override the effects of increased parathyroid hormone?

A. Vitamin D

B. Bone morphogenetic protein

C. Osteoprotegerin

D. Somatotrophin

E. Calcitonin

5. Osteoclasts would be found at which of the following sites in bone?

A. Howship's lacuna

B. Haversian canal

C. Canaliculi

D. Endosteum

E. Periosteum

6. In which of the following zones of endochondral ossifications would cells accumulate glycogen?

A. Resting zone

B. Zone of hypertrophy

C. Zone of ossification

D. Zone of proliferation

E. Zone of calcification

7. The differentiation of osteoclasts is controlled by which of the following factors?

    A. Calcitonin

    B. Parathyroid hormone

    C. Osteoprotegerin

    D. Osteocalcin

    E. Vitamin D

Bone of a 75-year-old man shows a decreased number of osteoblasts. However, the number of osteoclasts is greatly increased. This individual's bones are easily fractured.

8. Which of the following disease conditions would you expect?

    A. Osteoporosis

    B. Osteomalacia

    C. Osteopetrosis

    D. Osteochondrosis

    E. Rickets

9. Based on question 8, which of the following would best trigger differentiation of the osteoblasts?

    A. Parathyroid hormone

    B. Osteocalcin

    C. Macrophage-colony stimulating factor

    D. Core-binding factor (Cbfa-1)

    E. Osteoprotegerin

10. The hormone calcitonin acts directly on which of the following cell types?

    A. Osteocytes

    B. Osteoblasts

    C. Osteoclasts

    D. Chondroblasts

    E. Chondrocytes

11. At which of the following sites in bone would you find osteoblasts during active deposition of new bone matrix?

    A. Haversian canal

    B. Lacuna

    C. Periosteum

    D. Canaliculi

    E. Surface of bone

12. In which of the following bone sites are canaliculi found?

    A. Perichondrium

    B. Compact bone

C. Bone marrow

D. Newly mineralized bone matrix

E. Howship's lacuna

ANSWERS

1. The answer is B. Haversian systems contain a central haversian canal, interconnected by Volkmann's canals. Canaliculi contain processes of osteocytes, whereas lacunae house mature osteocytes.

2. The answer is E. Growth that increases the length of bone occurs at the epiphyseal plate, whereas increases in bone diameter occur at the diaphysis.

3. The answer is D. Parathyroid hormone stimulates osteoblasts to secrete osteoclast-stimulating factor. This factor induces osteoclasts to resorb bone to increase serum calcium levels.

4. The answer is E. Calcitonin acts directly on osteoclasts to suppress bone resorption.

5. The answer is A. Osteoclasts during the resorption process form a shallow depression in bone that is called Howship's lacunae.

6. The answer is B. During endochondral ossification, chondrocytes become highly vacuolated and accumulate glycogen in the zone of hypertrophy.

7. The answer is C. Osteoprotegerin is secreted by osteoblasts and acts to affect the differentiation of osteoclasts.

8. The answer is B. Osteomalacia is characterized by softening of bone because of a decrease in bone formation, whereas bone resorption is either normal or increased.

9. The answer is D. The factor that induces the differentiation of osteoblasts is core-binding factor (Cbfa-1). Parathyroid hormone activates osteoclasts and osteoprotegerin, and macrophage colony-stimulating factor (M-CSF) induces differentiation of osteoclasts. Osteocalcin binds calcium and is produced by osteocytes.

10. The answer is C. The only cell on which the pituitary hormone calcitonin acts directly is the osteoclast.

11. The answer is E. Osteoblasts depositing new bone matrix are found on the outer surface of bone.

12. The answer is B. Canaliculi connect osteocytes found within lacunae in compact bone.

CHAPTER 6
MUSCLE TISSUE

I. General Features

    A. Muscle tissue, composed primarily of muscle cells, also called fibers, is specialized for **contractility** and **conductivity.**

    B. Muscle is classified on a structural basis as **striated** (skeletal and cardiac), with regular transverse bands, or **smooth,** lacking transverse bands.

    C. Muscle is under either **voluntary** (skeletal) or **involuntary** (cardiac or smooth) control.

II. Skeletal Muscle

    A. Gross Organization
        1. Each individual muscle cell is surrounded by an **endomysium.**
        2. Masses of skeletal muscle fibers are arranged in regular bundles or **fascicles** enclosed by a **perimysium.**
        3. The **epimysium** is an external layer of connective tissue that surrounds fascicles of muscle fibers.

    B. Structure
        1. Skeletal muscle fibers measure 1–40 mm in length and 10–100 μm in width, and their cytoplasm, called **sarcoplasm,** contains thousands of myofibrils and large populations of mitochondria, called **sarcosomes** (Figure 6–1A and B).
        2. Skeletal muscle cells are multinucleated; nuclei are located at the **periphery** of a cylindrical fiber (Figure 6–1A).
        3. The longitudinally arranged myofibrils are composed of a series of **sarcomeres,** which is the unit of contraction, in striated muscle that consists of thin **actin** filaments and thicker **myosin** filaments (Figure 6–1A and B).
        4. The dark bands of sarcomeres are **A bands,** and at their center is a lighter zone called the **H band.** At the center of an H band is an **M line.** The light bands of a sarcomere are called **I bands,** and the dark line, the **Z line,** bisects **I bands** (Figure 6–2).
        5. A single sarcomere is delineated by 2 **Z lines,** containing 1 central A band and 2 I bands on either side of the A band.
        6. Myosin, a long macromolecule, has a head with an **adenosine triphosphatase (ATPase) active site** and **actin-binding site.** ATP binding to myosin activates it so that it may bind actin. Energy released by **ATPase** activity releases this complex. The thick filaments are primarily **myosin.**

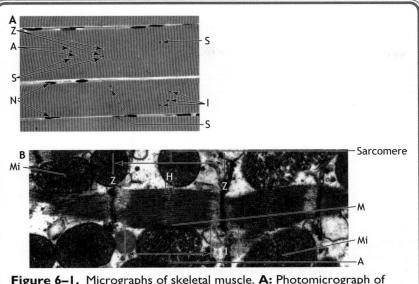

Figure 6–1. Micrographs of skeletal muscle. **A:** Photomicrograph of skeletal muscle. **B:** Electron micrograph of skeletal muscle. (N = nuclei; S = sarcomere; Z = Z line; I = I band; A = A band; Mi = mitochondria; H = H band; M = M line.)

7. The thin filaments that comprise I bands are made up of **actin, tropomyosin,** and 3 **troponin** peptides: TnT, TnI, and TnC.

8. **Actin** is a repeating structure (F-actin) composed of 2 strands in α-helical conformation. The strands are made up of individual globular units (G-actin) and contain myosin-binding sites.

9. **Tropomyosin,** a polypeptide, lies within the groove of the **actin α-helix** and covers the **myosin-binding sites.**

10. **TnT** binds to tropomyosin, **TnI** inhibits the binding of myosin heads to actin in resting muscle, and **TnC** binds **calcium** that triggers muscle contraction.

C. **Skeletal Muscle Contraction**

1. During muscle contraction, thin filaments slide past the thick filaments as a result of cross-links that form between actin and myosin.

2. Tension pulls the thin filament toward the thick filament.

3. The filaments maintain a constant length, but the H and I bands narrow, Z lines are pulled together, and the length of the A band remains the same.

4. The binding of Ca^{2+} to **TnC** causes a steric movement such that tropomyosin shifts out of the groove of the actin helix and thereby exposes the myosin-binding sites.

D. **T-Tubule System and Sarcoplasmic Reticulum**

1. The **T-tubule system** surrounds myofibrils and consists of invaginations of the sarcolemma at the **A–I junction.**

2. The T tubule conducts an impulse initiated at the sarcolemma throughout the muscle fiber system to coordinate muscle contraction.

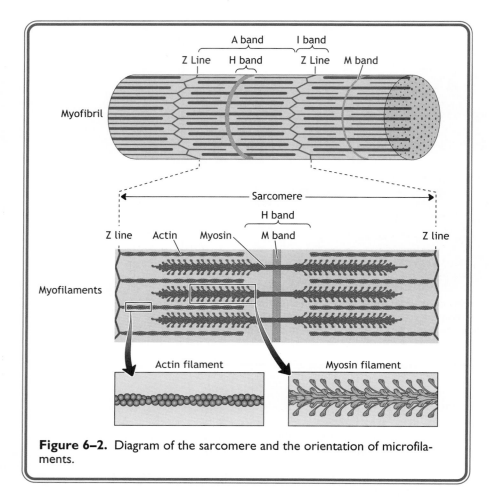

Figure 6–2. Diagram of the sarcomere and the orientation of microfilaments.

3. The **sarcoplasmic reticulum** is a modified smooth endoplasmic reticulum (SER) that consists of a network of membrane-limited flattened cisternae and sequesters and releases Ca^{2+} within the muscle fiber. Ca^{2+} release is triggered by signals from the T-tubule system.

4. The sarcoplasmic reticulum forms a triad with each tubule of the **T system,** and each sarcomere has 2 **triads.**

MUSCULAR DYSTROPHIES

• *Muscular dystrophies* *are several inherited diseases characterized by a loss of muscle fibers.*

• *One form of muscular dystrophy is a sex-linked, genetic disorder, affecting only males called* ***Duchenne's dystrophy.*** *These patients suffer from muscle wasting, which becomes evident by the age of 5. This disease results from a deletion or point mutation of the gene coding for* ***dystrophin.*** *This protein anchors the cytoskeleton of the* ***sarcolemma*** *to the* ***extracellular matrix.*** *This association allows these cells to withstand the rigors of muscle contraction.*

III. Cardiac Muscle

A. Structure

1. **Cardiac muscle fibers** have a single, central nucleus and a striated appearance (Figure 6–3A).
2. These cells are only 50–100 μm in length and bifurcate to differentiate them from skeletal muscle.
3. The **sarcoplasm** of cardiac muscle cells is more abundant, and **mitochondria** are more **numerous** than in skeletal muscle cells (Figure 6–3B).
4. **Glycogen** and **lipofuscin** are major components in the sarcoplasm of cardiac muscle cells.
5. **Intercalated disks** are formed at the Z line, which consists of a network of **desmosomes, zonulae adherens,** and **gap junctions.** Two proteins of intercalated disks are **vinculin** and α-**actinin.**
6. Modified cardiac fibers, called **Purkinje fibers,** are specialized for conducting electrical impulses within cardiac muscle. Cardiac muscle is not supplied by myoneural junctions; thus, Purkinje fibers are essential for coordinating cardiac muscle contraction.

B. T-Tubule System and Sarcoplasmic Reticulum

1. T tubules are located at the level of the Z lines, are described as **dyads,** and are larger than in skeletal muscle.
2. The sarcoplasmic reticulum of cardiac muscle fibers is not highly developed.

IV. Smooth Muscle

A. General Features

1. **Smooth muscle fibers** are nonstriated and spindle shaped, measuring 20–200 μm in length and varying in diameter from 3–9 μm (Figure 6–3C).
2. These cells have an elongated single central nucleus, a well-developed Golgi apparatus, but little sarcoplasmic reticulum.
3. This muscle is found primarily in the walls of the respiratory and gastrointestinal tracts and the walls of blood vessels, particularly large arteries.

B. Structure

1. The myofilaments actin and myosin in smooth muscle fibers are not arranged in an ordered fashion.
2. **Dense bodies,** also called **area densa,** are areas throughout the cytoplasm and along the inner surface of the sarcolemma that act as an actin attachment site and thus serve as **Z lines.**
3. An external basal lamina is found surrounding each smooth muscle cell.
4. Not all smooth muscle fibers receive a nerve terminal; thus, contraction of smooth muscle fibers is communicated by **gap junctions.**
5. Contraction of smooth muscle is triggered by an influx of Ca^{2+} that binds to **calmodulin.** This complex, in turn, binds to and activates a myosin kinase that phosphorylates myosin, which permits an interaction with actin.

V. Myoneural Junction

A. A single axon forms a close association with a single muscle fiber. This relationship is called a **motor unit** (Figure 6–4).

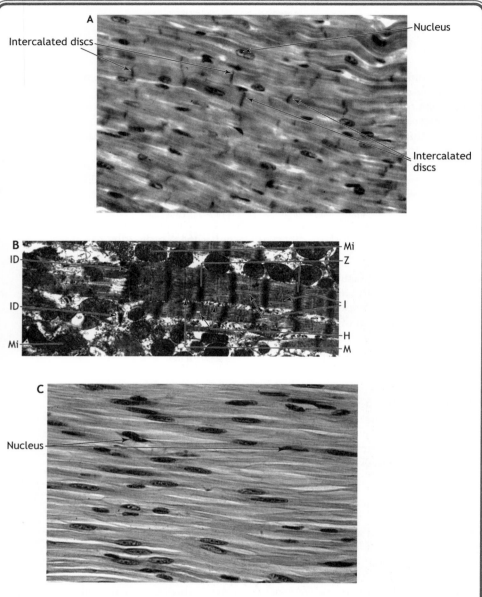

Figure 6–3. Micrographs of cardiac and smooth muscle. **A:** Photomicrograph of cardiac muscle. **B:** Electron micrograph of cardiac muscle. (ID = intercalated disk; Mi = mitochondrion; Z = Z line; S=sarcomere; M = M line; H = H band.) **C:** Photomicrograph of smooth muscle.

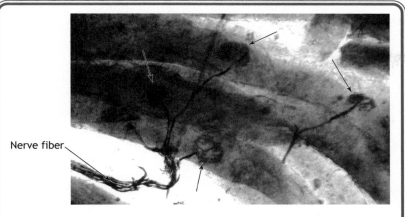

Figure 6–4. Photomicrograph of myoneural junctions. Arrows point to myoneural junctions.

B. When an axon approaches the muscle fiber, its myelin is lost, and several short terminals, or **boutons,** are located within a depression on the surface of the fiber. This structural complex is called a **myoneural junction** or **motor end plate.**

C. Synaptic vesicles within terminal branches of the axon contain **acetylcholine** (ACh), which is released via **exocytosis** triggered by action potentials.

D. The **sarcoplasmic reticulum** of muscle fibers is activated to release Ca^{2+}, which binds to TnC to initiate skeletal muscle contraction.

MYASTHENIA GRAVIS

- *Myasthenia gravis* is a genetic disorder in which **antibodies** to **acetylcholine** receptors are found in the neuromuscular junction. The muscular system will undergo fatigue and exhaustion.

- This disorder may involve a general muscle group such as muscles of respiration or a single muscle group.

CLINICAL PROBLEMS

1. In skeletal muscle fibers, which of the following would exhibit ATPase activity?

A. Actin filaments

B. Myosin filaments

C. Troponin

D. T-tubule system

E. Tropomyosin

2. Dense bodies within smooth muscle fibers most closely associate with which of the following components of skeletal muscle?

A. A band

B. I band

C. H band

D. M line

E. Z line

3. Which of the following best characterizes a cardiac muscle fiber?

A. Multiple nuclei

B. Lack of striations

C. Spindle-shaped fiber

D. Intercalated disks

E. Myoneural junctions

4. Which of the following events occurs during the process of contraction of skeletal muscle?

A. Release of calcium ions by sarcoplasmic reticulum

B. Calcium ions bind to tropomyosin

C. Actin filaments shorten

D. Z line disappears

E. Sarcomeres lengthen

5. Which of the following best characterizes a smooth muscle fiber?

A. Regularly arranged actin and myosin myofilaments

B. Calmodulin as the calcium ion-binding protein

C. Multiple, peripheral nuclei

D. Extensive myoneural junctions

E. Extensive sarcoplasmic reticulum

6. Which of the following regions contains the Z line in skeletal muscle?

A. H band

B. A band

C. I band

D. M band

E. At the junction of the A band and I band

A 5-year-old boy presents with wasting of muscles of the extremities. You suspect he is suffering from Duchenne's dystrophy.

7. Which of the following would be deficient or defective in this patient?

A. Actin

B. Myosin

C. Collagen

D. Dystrophin

E. Tropomyosin

8. Which of the following molecules is found within the groove created by actin filaments?

A. Troponin I

B. Myosin

C. Dystrophin

D. α-actinin

E. Tropomyosin

9. What is the function of the T-tubule system in skeletal muscle?

A. Regulates Ca^{2+} release from sarcoplasmic reticulum

B. Stores Ca^{2+}

C. Provides channel for Ca^{2+} movement with muscle

D. Wraps individual muscle fibers to form a fascicle

E. Site of attachment of actin filaments

You are examining a heart muscle tissue section with an electron microscope. You notice a cluster of cells intermixed with cardiac muscle fibers that have what appears to be glycogen within their cytoplasm.

10. Which of the following are you describing?

A. Fibroblasts

B. Purkinje cells

C. Smooth muscle fibers

D. Endothelial cells

E. Nerve axons

ANSWERS

1. The answer is B. ATPase activity is found on the globular head of myosin filaments.

2. The answer is E. Dense bodies within smooth muscle function to anchor actin and other intermediate filaments and thus serve as a Z line.

3. The answer is D. Intercalated disks are only found in cardiac muscle. This junctional complex functions to allow cardiac muscle to communicate. Cardiac muscle does not have myoneural junctions, but its contraction is coordinated by the sinoatrial (SA) and atrioventricular (AV) nodes and Purkinje fibers.

4. The answer is A. During contraction of skeletal muscle, calcium ions are released from the sarcoplasmic reticulum and bind to troponin C. Sarcomeres shorten during contraction, and the Z line does not disappear during muscle contraction.

5. The answer is B. The calcium-binding protein in smooth muscle is calmodulin, whereas troponin C binds calcium in skeletal and cardiac muscle. Smooth muscle fibers have single, central nuclei and little sarcoplasmic reticulum.

6. The answer is C. The Z line in skeletal muscle is found within an I band. The I band stretches across 2 adjacent sarcomeres.

7. The answer is D. Dystrophin is either defective or deficient in Duchenne's muscular dystrophy patients.

8. The answer is E. Tropomyosin is located within the groove created by the actin filament.

9. The answer is A. The T-tubule system sends a signal received at the sarcolemma. This signal is transported to the sarcoplasmic reticulum, which responds by releasing Ca^{2+}, which binds to troponin C, triggering a muscle contraction event.

10. The answer is B. The cells within the heart that contain glycogen are Purkinje cells (or fibers). These cells are modified cardiac muscle cells that transmit the signal to coordinate muscle contraction.

CHAPTER 7
PERIPHERAL BLOOD

I. Components of Blood

A. The **plasma** component of blood consists of albumin, globulins, and fibrinogen.

B. The **cellular** constituent of blood includes erythrocytes, leukocytes, and platelets.

II. Hematocrit

A. A **hematocrit** measures the volume of packed red blood cells per unit volume of total blood after centrifugation.

B. At the interface between packed **erythrocytes** (45%) and **plasma** (55%), a layer of leukocytes (white blood cells) is observed, called a **buffy coat** (1%).

C. Hematocrit values vary depending on age, sex, environment, and health.

III. Erythrocytes

A. Mature erythrocytes (Figure 7–1) are **biconcave disks,** 7.5 μm in diameter and 2 μm thick, and are flexible bags of hemoglobin because organelles and nuclei are no longer present.

B. Adult males have 5.5 million/mm^3 erythrocytes in the blood and **adult females have 4.5 million/mm^3.**

C. The life span of an erythrocyte is about **120 days.** Erythrocytes damaged by passage through the circulatory system are removed within the **spleen** and **bone marrow** by **macrophages.**

D. Reticulocytes are immature erythrocytes with small levels of ribosomal RNA (rRNA). These cells make up 1–2% of circulating red blood cells in a healthy individual and **represent a clinical measure of the health of a person.**

E. Hemoglobin is composed of 2 α chains and 2 β chains and 4 heme groups, which bind oxygen (O$_2$).

ANISOCYTOSIS

- **Anisocytosis** is a condition in which a large population of erythrocytes have varied diameters. **Macrocytes** are erythrocytes larger than 9 μm in diameter, and **microcytes** are smaller than 6 μm in diameter.

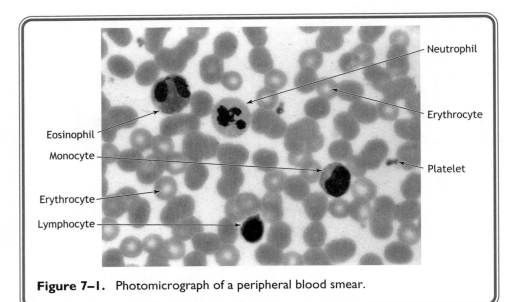

Figure 7–1. Photomicrograph of a peripheral blood smear.

ANEMIA

- *Anemia* is a decreased concentration of **hemoglobin** in blood that can result from loss of blood (**hemorrhage**), reduced production of red blood cells, red blood cells that contain insufficient hemoglobin (iron deficiency anemia), and accelerated red blood cell destruction.
- *Pernicious anemia* results from decreased secretion of **intrinsic factor** by cells of the intestinal mucosa. This factor is required for absorption of **vitamin B$_{12}$ from the diet. The reduced vitamin B^{12}** levels result in decreased erythrocyte production.
- *Sickle cell anemia* is an inherited disorder involving a point mutation of the β-**globin gene,** resulting in an amino acid substitution in the hemoglobulin molecule (HbS). The **HbS** protein polymerizes and aggregates within the cytoplasm of the erythrocyte, which alters its shape. Formation of HbS results in **chronic hemolytic anemia** and blockage of small blood vessels.

THALASSEMIA

- *Deficient synthesis of the* α *or* β *chain of hemoglobin* leads to α-**thalassemia** or β-**thalassemia,** respectively. These disorders result in severe to asymptomatic anemia.

POLYCYTHEMIA

- *Absolute polycythemia* is marked by excessive levels of erythrocytes caused by an overactive production of these cells within the bone marrow.
- *Relative polycythemia* results from a decreased plasma level without increased erythrocyte production.

IV. Leukocytes

    A. Leukocytes are involved in cellular and humoral (antibody mediated) defense against foreign material (Figures 7–1, 7–2).

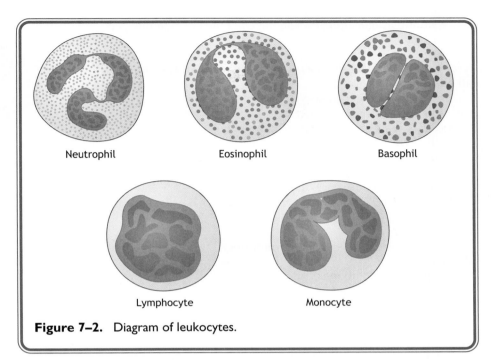

Neutrophil   Eosinophil   Basophil

Lymphocyte   Monocyte

Figure 7–2. Diagram of leukocytes.

B. **Granulocytes** are polymorphonuclear leukocytes, 12–15 μm in diameter (Figures 7–1, 7–2). Three types of granulocytes are neutrophils, eosinophils, and basophils.
1. Neutrophils
   a. The nucleus of a **neutrophil** has 2–5 lobes linked by fine threads of chromatin. Neutrophils with 6-lobed nuclei are typically found in the aged population.
   b. The specific granules within the cytoplasm do not stain, resulting in a pale cytoplasm.
   c. The **azurophilic granules** are primary **lysosomes.**
   d. The primary function of neutrophils is **phagocytosis** of bacteria.
   e. These cells comprise **50–70%** of circulating leukocytes in adults.
2. Eosinophils
   a. The nucleus of an **eosinophil** is bilobed.
   b. The specific granules within the cytoplasm stain **reddish orange.**
   c. These cells function to combat **parasitic invasions** and release proteins such as **major basic protein,** eosinophil cationic protein, and eosinophil peroxidase.
   d. Eosinophils are **2–4%** of circulating leukocytes in adults.
3. Basophils
   a. The nucleus of a **basophil** is kidney shaped but obscured by overlying deep purple-stained granules.
   b. These granules contain **heparin** and **histamine.**
   c. The primary function of basophils is in **immediate hypersensitivity reactions.** The **immunoglobulin E (IgE) receptor** is expressed on the surface of the basophil, like mast cells.
   d. Basophils are **< 1%** of circulating leukocytes in adults.

C. **Agranulocytes are mononuclear leukocytes** and are composed of **lymphocytes** and **monocytes.**

1. Lymphocytes

   a. **Lymphocytes have a compact nucleus, sometimes with an indentation, and a small rim of cytoplasm.**

   b. These cells comprise **20–40%** of leukocytes in adults.

   c. Most lymphocytes are 6–8 μm in diameter.

   d. The precursor cells to lymphocytes originate in **bone marrow.**

   e. B lymphocytes (B cells)

     (1) **B cells** are important for **humoral immunity,** which depends on circulating antibodies. Antigen-stimulated B cells differentiate into plasma cells, which secrete antibodies.

     (2) Differentiation of B cells occurs in bone marrow and B cells make up **15%** of lymphocytes.

   f. T lymphocytes (T cells)

     (1) **T cells** are important for **cell-mediated immunity,** such as rejection of **transplanted organs.**

     (2) T cells differentiate in the **thymus** and comprise **80%** of lymphocytes.

     (3) These cells secrete lymphokines.

     (4) Types of T cells include T helper, suppressor T cells, and cytotoxic T cells.

   g. Null cells

     (1) Null cells represent 5% of lymphocytes.

     (2) These cells are **natural killer (NK) cells,** which kill aberrant cells without T cells, and **stem cells,** which give rise to blood elements.

2. Monocytes

   a. The nucleus of **monocytes** is oval or kidney shaped and is generally located eccentrically.

   b. These cells are 12–20 μm in diameter.

   c. Monocytes originate within the bone marrow but differentiate into **macrophages** after migrating into connective tissue, such as alveolar macrophages in the lung and Kupffer's cells in the liver.

   d. Monocytes comprise **3–8%** of circulating leukocytes in the adult.

GRANULOCYTE CANCERS

- *Leukemias,* which are a progressive proliferation of leukocytes, are classified according to the major cell type and the period of duration. This blood disorder can be *acute,* lasting a few months, or *chronic,* lasting over 1 year, and may result in enlargement of the liver, spleen, and lymph nodes.

- *Myelogenous leukemia* is a malignancy of granulocyte precursor cells, whereas *lymphocytic leukemia* is cancer of lymphocyte precursor cells.

V. Platelets

A. Platelets are cellular structures, amounting to about 200,000/mm³ blood, that originate from fragmentation of **megakaryocytes** within the bone marrow. The megakaryocyte is derived from a **megakaryocytoblast.**

B. **Thrombopoietin** stimulates the maturation of megakaryocytes, giving rise to platelets.

C. Platelets are fragments, 2–4 μm in diameter, and consist of a peripheral **hyalomere,** a central **granulomere,** open canalicular system, and various granules.

D. The products of platelets, such as platelet factor IV, **von Willebrand factor,** thrombospondin, and platelet-derived growth factor (PDGF), promote blood clotting and repair of gaps in blood vessel walls.

E. A blood clot consists of aggregating platelets and fibrils formed from fibrin.

THROMBOCYTOPENIA

- **Thrombocytopenia** is a disorder marked by a reduced level of circulating platelets.
- **Thrombocytopenic purpura** is a chronic autoimmune disease in which antibodies to platelets interfere with their blood-clotting function.

HEMOPHILIA

- **Hemophilia A** is a sex-linked inherited disorder in which a clotting factor, **factor VIII,** is reduced in amount or activity.
- **Hemophilia B** is a sex-linked inherited disease in which **factor IX** is nonfunctional or deficient.

CLINICAL PROBLEMS

Examination of a normal peripheral blood smear reveals a cell more than twice the diameter of an erythrocyte with a kidney-shaped nucleus. These cells were < 10% of total leukocytes.

1. Which of the following cell types is being described?

    A. Monocyte

    B. Basophil

    C. Eosinophil

    D. Neutrophil

    E. Lymphocyte

Your patient is suffering from an infestation of parasites. To determine the proper therapy, you perform an analysis of a peripheral blood sample.

2. Which of the following cells would you expect to be increased in a peripheral blood smear?

    A. Lymphocytes

    B. Basophils

    C. Neutrophils

    D. Monocytes

    E. Eosinophils

3. A decrease in the number of tissue macrophages would result from reduced activity of precursor cells to which of the following cell types?

   A. Basophils

   B. Erythrocytes

   C. Monocytes

   D. Neutrophils

   E. Megakaryocytes

4. Which of the following cells is the primary phagocytic cell within peripheral blood?

   A. Eosinophil

   B. Monocyte

   C. Basophil

   D. Neutrophil

   E. Lymphocyte

5. Pernicious anemia develops as a consequence of which of the following?

   A. Amino acid change of the globin molecule

   B. Decreased levels of intrinsic factor

   C. Decreased synthesis of hemoglobins

   D. Decreased platelets

   E. Decreased iron

While examining a peripheral blood smear, you notice an increased number of cells slightly larger than an erythrocyte. These cells have a large, round nucleus and scant cytoplasm.

6. Which of the following would cause this condition?

   A. Bacterial infection

   B. Parasitic infection

   C. Organ transplant rejection

   D. Pollen infestation

   E. Hemorrhage

A young man complains that he bruises easily. A laboratory report showed a normal white cell count. Erythrocytes were 5 million/mm^3, whereas platelets were 50,000/mm^3.

7. Which of the following was the diagnosis?

   A. Polycythemia

   B. Thrombocytopenia

   C. Thalassemia

   D. Hemophilia

   E. Pernicious anemia

8. Which of the following blood cells differentiate outside of the bone marrow?
   A. T lymphocytes
   B. Neutrophils
   C. Basophils
   D. Eosinophils
   E. Megakaryocytes

9. An analysis of a peripheral blood smear showed that reticulocytes were 10% of total cells. Which of the following represents the proper diagnosis?
   A. Humoral-mediated immune response
   B. Myelocytic leukemia
   C. Infection
   D. Cell-mediated immune response
   E. Hemorrhage

10. Which of the following is not derived by a mitotic division of a precursor cell?
    A. Proerythroblast
    B. Monocyte
    C. Polychromatophilic erythroblast
    D. Platelet
    E. Osteoblast

Examining a peripheral blood smear, you note a cell type that consists of both large and small cells intermixed with normal size cells, a conditioned described as anisocytosis.

11. This abnormal condition refers to which of the following?
    A. Monocytes
    B. Lymphocytes
    C. Erythrocytes
    D. Basophils
    E. Platelets

ANSWERS

1. The answer is A. The monocyte is the only nucleated cell in a normal peripheral blood smear that has a diameter twice that of an erythrocyte. Large lymphocytes have a round, not a kidney-shaped, nucleus.

2. The answer is E. The blood cell that destroys parasites is the eosinophil. The factors released by eosinophils kill the parasite-forming pores in its wall by engulfing antigen-antibody complexes.

3. The answer is C. Macrophages develop from monocytes that migrate from the blood into connective tissue.

4. The answer is D. The primary function of neutrophils in blood is phagocytosis of bacteria and other microorganisms.

5. The answer is B. Pernicious anemia occurs when decreased intrinsic factor is released into the gastrointestinal tract, resulting in decreased vitamin B_{12} absorption. Reduced vitamin B_{12} affects decreased erythrocyte production.

6. The answer is C. The cell in peripheral blood that is only slightly greater in diameter than an erythrocyte and has a round nucleus and little cytoplasm is a lymphocyte. Lymphocytes, specifically T cells, function directly or indirectly in response to organ transplants.

7. The answer is B. Thrombocytopenia is characterized by a reduction in the number of platelets. Platelets normally number between $2–4 \times 10^5/mm^3$; the reduced platelet count would indicate thrombocytopenia.

8. The answer is A. T lymphocytes or T cells differentiate within the thymus. All of the other cells differentiate within the bone marrow.

9. The answer is E. A hemorrhage would stimulate a rapid increase in erythrocyte production, which would be observed as increased numbers of reticulocytes above the normal 1% of total cells.

10. The answer is D. The only cell product not derived from a mitotic division of a precursor cells is a platelet. Platelets are fragments of a megakaryocyte formed within the bone marrow.

11. The answer is C. The condition in which erythrocytes exhibit larger and smaller than normal diameters is referred to as anisocytosis.

CHAPTER 8
HEMATOPOIESIS

I. Hematopoiesis and Bone Marrow

- **A. Hematopoiesis** is the process of blood cell formation in bone marrow.
- **B.** Bone marrow is found in **medullary canals** of long bones and in the **cavities** of cancellous bones (Figure 8–1).
- **C. Active-red** and **adipose-filled yellow** are the 2 types of bone marrow.
    - **1. Red bone marrow** is composed of a stroma, hematopoietic cords, and sinusoids. The main **functions** of red bone marrow are production of blood cells, destruction of malformed erythrocytes, and storage of iron derived from hemoglobin breakdown by macrophages.
    - **2. Yellow bone marrow** acts as a storage organ for fat and a reserve site of hematopoietic tissue. Under proper stimulation, yellow bone marrow can transform into red marrow.

II. Blood Cell Differentiation

- **A.** All blood cells originate from undifferentiated **mesenchymal** (stromal) **cells** within the bone marrow. These cells, when stimulated by **stem cell factor** (SCF), give rise to **pluripotent hematopoietic stem cells (PHSCs).** SCF is also called **c-kit ligand.**
- **B. PHSCs** give rise to **myeloid** and **lymphoid stem cells,** which have a limited potential to a single-cell fate but are capable of self-renewal.
- **C. Myeloid stem cells** give rise to **erythroid, eosinophil, basophil, granulocyte-macrophage (GM),** and **megakaryocyte colony-forming units (CFUs).** The CFUs under the influence of specific factors are driven to undergo further differentiation to a mature cell fate.
- **D.** Early hematopoietic stem cells are controlled by **homeobox genes.** *Hox1* controls the myeloid series, except the erythroid series, which is controlled by the *Hox2* gene.

III. Erythropoiesis

- **A. Erythropoiesis** (Figure 8–2A) is the maturation of red blood cells. The renal hormone **erythropoietin** controls red blood cell production.
- **B.** The first cell of the erythrocytic series is the **proerythroblast,** which develops from an **erythroid CFU.** The proerythroblast has a large nucleus and a thin rim of basophilic cytoplasm because of the presence of polyribosomes.

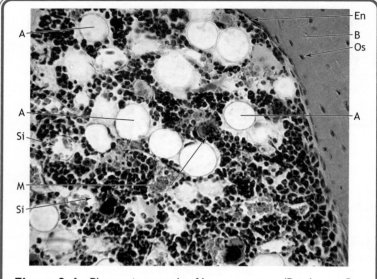

Figure 8–1. Photomicrograph of bone marrow. (B = bone; Os = osteocyte; A = adipose cell; M = megakaryocyte; Si = sinusoid; En = endosteum.)

C. The immediate progeny of the proerythroblast is the **basophilic erythroblast** (Figure 8–2A), which is somewhat smaller with a basophilic cytoplasm. **Hemoglobin** is first observed in this cell type.

D. Further mitosis leads to a smaller **polychromatophilic erythroblast.** The nucleus of this cell is increasingly heterochromatic, and the cytoplasm is somewhat bluish gray as a result of increasing synthesis of **hemoglobin** (Figure 8–2B).

E. The **orthochromatophilic erythroblast** results from division of the polychromatophilic erythroblast. This cell has a small, dense nucleus and a pinkish cytoplasm as **hemoglobin** synthesis continues. This cell is not capable of mitosis. At the end of this stage, its **nucleus** is extruded (Figure 8–2B).

F. A **reticulocyte** is an immature erythrocyte that has a few remaining polyribosomes, stained with supravital dyes.

G. The mature **erythrocyte** is formed after loss of the polyribosomes.

ERYTHROBLASTOSIS FETALIS

• *Erythroblastosis fetalis is a hemolytic anemia resulting from an **anti-Rh antibody** produced in the mother in response to the Rh factor of the fetus.*

• *This disorder is characterized by erythroblasts within peripheral blood of the fetus and enlargement of the spleen and liver.*

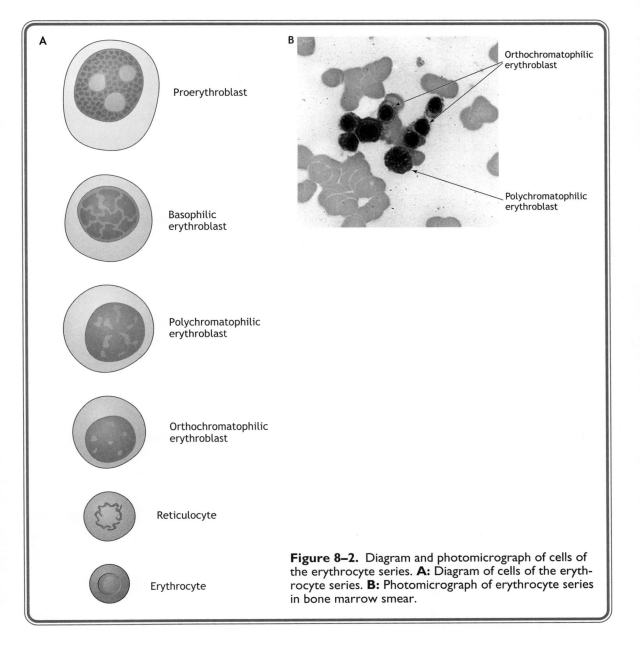

Figure 8–2. Diagram and photomicrograph of cells of the erythrocyte series. **A:** Diagram of cells of the erythrocyte series. **B:** Photomicrograph of erythrocyte series in bone marrow smear.

ERYTHROLEUKOSIS

- *Erythroleukosis* is a condition in which cells of the erythroid series, in addition to leukocytic cells, undergo uncontrolled proliferation.

MEGALOBLASTIC ANEMIA

- *Megaloblastic anemia* is characterized by larger than normal erythrocytes in the peripheral blood. Bone marrow is populated by large numbers of erythroblasts in the bone marrow.
- In severe conditions, nucleated erythrocytes may be found in peripheral blood, and the **reticulocyte** levels are below normal.

IV. Granulopoiesis

    A. In the granulocytic series, the **myeloblast,** the most immature, recognizable cell in the myeloid series, has no granules and a light blue cytoplasm. The nucleus has 2–3 nucleoli. This cell is derived from a **CFU** under the influence of specific **colony-stimulating factors (CSFs)** (Figure 8–3A).

        1. The **GM-CFU** is driven by GM-CSF, granulocyte CSF (G-CSF), and interleukin (IL)-3 to the neutrophil or monocyte series.

        2. The **eosinophil CFU** is driven by GM-CSF, IL-3, and IL-5 to the eosinophilic series.

        3. The **basophil CFU** is driven by IL-3 and IL-4 to the basophilic series.

    B. The next cell in the lineage is the **promyelocyte.** The cytoplasm is light blue, and its nucleus may have a single nucleolus. This cell gives rise to the **3 types of granulocytes.**

    C. The **myelocyte** shows the first sign of cellular differentiation in the granulocyte series by the appearance of **specific cytoplasmic granules.** This cell is potentially mitotic (Figure 8–3B).

    D. **Nuclear indentation,** observed in the **metamyelocyte** stage, indicates the beginning of nuclear lobe formation. Metamyelocytes are not capable of division.

    E. **Neutrophils,** and possibly eosinophils and basophils, have an intermediary **band** or **stab cell stage.** The appearance of stab cells in blood is of clinical significance.

MYELOBLASTIC LEUKEMIA

- *Acute myeloblastic leukemia* is derived from transformed stem cells that do not undergo maturation beyond the **myeloblast stage.** Excessive numbers of myeloblasts are found in circulating blood and organs.
- The therapy for this leukemia involves chemotherapy followed by bone marrow transplantation.

MYELOGENOUS LEUKEMIA

- *Chronic myelogenous leukemia* is characterized by uncontrolled proliferation of cells of the granulocytic series, usually the neutrophil.
- Increased cell numbers are observed within bone marrow, peripheral blood, organs, and tissues.

GRANULOCYTIC LEUKOCYTOSIS

- *Granulocytic leukocytosis* is characterized by an increase in neutrophils, eosinophils, or, rarely, basophils within the peripheral blood. Leukocytes can increase to as high as $1 \times 10^5/\mu L$.
- This condition can result from bacterial infections (neutrophilic) or allergies (eosinophilic).

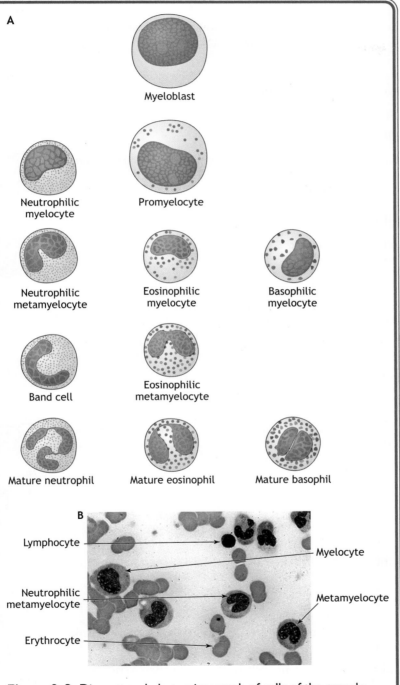

Figure 8–3. Diagram and photomicrograph of cells of the granulocyte series. **A:** Diagram of granulocyte series. **B:** Photomicrograph of granulocyte series in bone marrow smear.

V. Lymphopoiesis

A. Progenitor cells of lymphocytes originate in bone marrow as large **lymphoid stem cells.** These stem cells are driven by stem cell factor (SCF) and IL-7 to **prolymphocytes.**

B. The **prolymphocytes** are driven by IL-7 to pro-**T cells,** which differentiate within the thymus. The **flt3-ligand** drives production of pro-**B cells,** which differentiate in bone marrow.

LYMPHOBLASTIC LEUKEMIA

- *Acute lymphoblastic leukemia* (ALL) is composed of blast cells of the lymphocytic lineage, both pre-B and pre-T cells.
- *This leukemia usually affects children.*

LYMPHOBLASTIC LEUKEMIA

- *Chronic lymphoblastic leukemia* (CLL) is composed of **B cells.**
- *It usually affects older adults.*

VI. Monopoiesis

A. The **monoblast** is the progenitor cell of monocytes found within bone marrow. This cell is derived from the **GM-CFU** driven by GM-CSF and IL-3.

B. Further differentiation leads to the **promonocyte,** which divides twice to form **monocytes.**

C. **Monocytes** mature into **macrophages** after migration into connective tissues.

MONOCYTIC LEUKEMIA

- *Monocytic leukemia* consists of large numbers of cells of the monocytic series. These cells can accumulate within the peripheral blood.

MONOCYTOSIS

- *Monocytosis* is characterized by an increase in the number of monocytes in peripheral blood.
- *This blood disorder is caused by malaria, tuberculosis, and systemic lupus erythematosus.*

CLINICAL PROBLEMS

Your patient suffers from a blood disorder that requires chemotherapy. A bone marrow transplant is subsequently performed.

1. Transplants containing which of the following cells in the erythrocytic series would not provide proliferative cells?

   A. Proerythroblast

   B. Orthochromatophilic erythroblast

C. Polychromatophilic erythroblast

    D. Basophilic erythroblast

    E. Hematopoietic stem cell

2. At which of the following stages are specific cells of the granulocytic series first recognizable?

    A. Myelocyte

    B. Hematopoietic stem cell

    C. Promyelocyte

    D. Metamyelocyte

    E. Myeloblast

3. Hemoglobin is first seen in which of the following cells of the erythrocytic series?

    A. Orthochromatophilic erythroblast

    B. Polychromatophilic erythroblast

    C. Reticulocyte

    D. Basophilic erythroblast

    E. Proerythroblast

4. In which of the following cells of the erythrocytic series is the nucleus extruded?

    A. Reticulocyte

    B. Proerythroblast

    C. Basophilic erythroblast

    D. Polychromatophilic erythroblast

    E. Orthochromatophilic erythroblast

While examining cells of a bone marrow smear with a light microscope, you notice a cell whose cytoplasm has no noticeable granules. The nucleus of this cell is indented and contains no nucleoli.

5. Which of the following cells are you examining?

    A. Polychromatophilic erythroblast

    B. Neutrophilic metamyelocyte

    C. Basophilic erythroblast

    D. Promyelocyte

    E. Myelocyte

6. Which of the following best characterizes bone marrow?

    A. Yellow bone marrow does not function to produce blood cells

    B. Lymphocytes mature in bone marrow

    C. Destruction of erythrocytes is carried out in red bone marrow

    D. Monocytes mature in the bone marrow

    E. All blood cells undergo maturation within the bone marrow

A sample of peripheral blood from a patient reveals a larger than normal population of cells that have large, round nuclei with 1 or 2 nucleoli. The cytoplasm of these cells shows azurophilic granules.

7. Which of the following forms of leukemia would you suspect?

   A. Promyelocytic leukemia

   B. Basophilic leukemia

   C. Lymphoblastic leukemia

   D. Stem cell leukemia

   E. Eosinophilic leukemia

A peripheral blood smear of a person suffering from a mild form of leukemia showed a large number of cells with more than 1 nucleolus.

8. Which of the following cells is being described?

   A. Neutrophilic myelocyte

   B. Orthochromatophilic erythroblast

   C. Neutrophilic metamyelocyte

   D. Polychromatophilic erythroblast

   E. Myeloblast

Your patient suffers from lymphocytic leukemia. Chemotherapy is administered to destroy cancerous cells of the bone marrow. Precursor cells to erythrocytes are destroyed.

9. To specifically re-establish the erythrocytic cell line, which of the following cells should be transplanted?

   A. Reticulocytes

   B. Orthochromatophilic erythroblasts

   C. Megakaryoblasts

   D. Basophilic erythroblasts

   E. Metamyelocytes

10. As cells undergo the process of erythropoiesis, which of the following occurs?

   A. Nuclei become lobulated

   B. Nuclei decrease in size

   C. Nucleoli increase in number

   D. Rough endoplasmic reticulum (RER) increases within the cytoplasm

   E. Chromatin becomes more diffuse

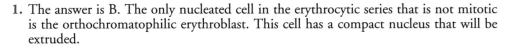

ANSWERS

1. The answer is B. The only nucleated cell in the erythrocytic series that is not mitotic is the orthochromatophilic erythroblast. This cell has a compact nucleus that will be extruded.

2. The answer is A. In cells of the granulocytic series, specific granules first begin to be deposited in the cytoplasm of myelocytes. This is the stage at which one can first distinguish cell lineage for eosinophils, basophils, and neutrophils.

3. The answer is D. Hemoglobin is first synthesized in basophilic erythroblasts. The cytoplasm of these cells is slightly pink compared with the proerythroblast.

4. The answer is E. During the process of maturation of erythrocytes, hemoglobin is manufactured and the nucleus becomes condensed. The condensed nucleus is extruded from the orthochromatophilic erythroblast.

5. The answer is B. In granulocytic cells of bone marrow, indentation of the nucleus is a gradual event, culminating in its lobulation. The cell without specific granules and an indented nucleus would be identified as a neutrophilic metamyelocyte.

6. The answer is C. In addition to the production of blood cell precursors, bone marrow also provides a site for the destruction of erythrocytes, which may be defective morphologically, by the action of macrophages.

7. The answer is A. The presence of azurophilic granules within the cytoplasm of the described cells indicates the neutrophilic lineage. Thus, the only cell that would have these granules would be a promyelocyte, which represents the precursor cell of neutrophils.

8. The answer is E. The myeloblast is the only cell of those described that has more than 1 nucleolus. The orthochromatophilic and polychromatophilic erythroblasts have condensed nuclei, and thus no nucleoli, whereas neutrophilic myelocytes and promyelocytes may have a single nucleolus.

9. The answer is D. The basophilic erythroblast is mitotically active and is a precursor to the polychromatophilic erythroblast. The reticulocyte does not have a nucleus, whereas the other cells, although potentially mitotic, are normally precursor cells of the myeloid series or the megakaryocyte.

10. The answer is B. As erythrocytes continue to differentiate, the nucleus becomes smaller and the chromatin more condensed. Heterochromatin becomes more prominent, not euchromatin. The RER and nucleoli become less prominent in cells as differentiation continues. Lobulation is distinctive of granulocytic differentiation, particularly of neutrophils.

CHAPTER 9
CENTRAL NERVOUS SYSTEM

I. General Features

A. **Nervous tissue** is distributed throughout the body to form an integrated communications network.

B. The **functions** of nervous tissues are to detect, analyze, integrate, and transmit information generated by sensory stimuli and chemical or mechanical changes and to organize and coordinate functions of the body.

C. The **central nervous system** (CNS) consists of the **brain** (cerebrum, cerebellum, brainstem) and the **spinal cord,** which are surrounded by the **meninges.**

II. Neurons

A. General Features
1. **Neurons,** the largest cells within the CNS, respond to stimuli by altering electrical potential across membranes at nodes of Ranvier and synapses.
2. When the difference of an electrical potential is spread throughout the membrane, an **action potential,** or nerve impulse, results.

B. Types of Neurons
1. **Multipolar neurons** have more than 2 dendrites and a single axon. These neurons are found throughout the CNS and autonomic ganglia (Figure 9–1).
2. **Bipolar neurons** have a single axon and a dendrite. These cells are found in the retina and olfactory epithelium (Figure 9–1).
3. **Pseudounipolar neurons** have a single process, which divides into 2 branches to form a T-shaped structure. These neurons are found in cranial and spinal ganglia (Figure 9–1).

C. Perikarya
1. **Perikarya** (soma or neuron cell bodies) are the trophic center of cells but can also be receptive to stimuli. Perikarya range from 4–5 μm to as large as 150 μm in diameter (Figure 9–2A).
2. The perikaryon contains a nucleus and cytoplasm but does not include the processes, and it receives **nerve endings** that transmit excitatory or inhibitory stimuli from other neurons.
3. The nucleus is usually large and contains mostly **euchromatin** and a prominent nucleolus.
4. In most neurons, the rough endoplasmic reticulum (RER), called **Nissl bodies,** is highly developed.

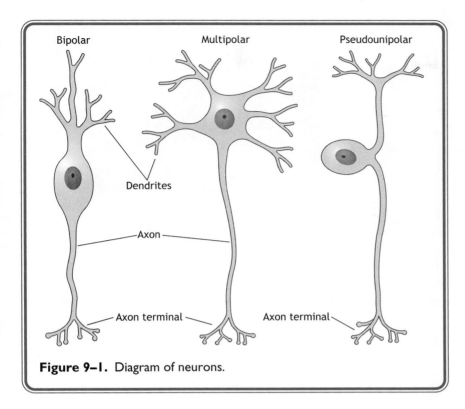

Figure 9–1. Diagram of neurons.

5. The **Golgi complex** is found only in the perikaryon, whereas **mitochondria** are also found in axon terminals.
6. Filaments
   a. **Intermediate filaments** having a diameter of about 10 nm are found in the perikaryon and processes.
   b. The perikaryon also contains **microtubules,** which can be associated with microfilaments.
7. Granules
   a. **Lipofuscin,** which appears light brown, is sometimes referred to as **age pigment.** These granules accumulate in some neuron cell bodies during aging.
   b. **Melanin granules,** which appear dark brown or black, are also found in perikarya of neurons, especially in the substantia nigra of the midbrain.

D. **Dendrites**
   1. Most neurons have several **dendrites** extending from their perikaryon. These structures greatly increase the receptive capacity of the neuron.
   2. RER and free ribosomes are found in dendrites, although the Golgi apparatus is usually not present.
   3. Neurofilaments and microtubules are aligned along the long axis of the dendrite.

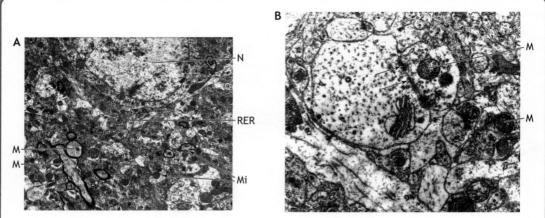

Figure 9–2. Electron micrographs of central nervous system tissue. **A:** Perikaryon. (N = nucleus; Mi = mitochondrion; RER = rough endoplasmic reticulum; M = myelinated axon.) **B:** Synapse. Arrows point to synaptic region. (M = mitochondria.)

4. Dendritic microtubules aid in the transport of macromolecules to distal regions.
5. Dendrites receive synaptic contacts at points along their length, and **dendritic spines,** which are small protrusions, may also receive synaptic contacts.

E. Axons
1. The plasma membrane of the axon is called the **axolemma**; its cytoplasm is referred to as **axoplasm.** This distinction is made because the components of the plasma membrane and cytoplasm of the axon differ from those of the neuron perikarya.
2. The **axon hillock** is a conical elevation from which the axon originates from the perikaryon.
3. In contrast to dendrites, the axon hillock is devoid of RER and free ribosomes but contains mitochondria.
4. Many **microtubules** and **neurofilaments** are often arranged in bundles and function in the movement of membrane vesicles.
5. That part of the axon between the axon hillock and the point at which the axon becomes myelinated is called the **initial segment.**

F. Synapses
1. The site of a junction of an axon with a dendrite, perikaryon, or other axon is called a **synapse** (Figure 9–2B).
2. The region of extracellular space between the presynaptic and postsynaptic membranes is called the **synaptic cleft.** This cleft is not a free space but is occupied by glial processes and contains mucopolysaccharides and proteins. These macromolecules function to limit transmitter diffusion and transport neurotransmitters.
3. The cytoplasm of the terminal endings contains numerous **synaptic vesicles** (20–60 nm), which are composed of chemical substances called **neurotrans-**

mitters. These neurotransmitters are responsible for the transmission of a nerve impulse across the synapse.

4. The neurotransmitter **acetylcholine** is found in round, clear vesicles, whereas vesicles containing **norepinephrine** are 40–60 nm in diameter with a dense-staining core.

5. Neurotransmitters are released at the **presynaptic membrane** by **exocytosis** and bind to specific **receptors** on the **postsynaptic membrane** to initiate an excitatory or inhibitory response.

6. Synaptic vesicles fuse with the presynaptic membranes and undergo **endocytosis** and recycle to form new synaptic vesicles.

PARKINSON'S DISEASE

- *Parkinson's disease results from the loss of neurons in the substantia nigra and dopamine levels in the corpus striatum.*
- *This disease is characterized by tremors and muscle weakness that usually develop late in life. Parkinson's disease may exhibit autosomal dominant inheritance.*

HUNTINGTON'S DISEASE

- *Huntington's disease is an autosomal dominant disease characterized by degeneration of striatal neurons, particularly of the caudate nucleus, leading to dementia and movement disorders. The mutant **huntingin protein** coded by the **HD** gene accumulates within intranuclear inclusions.*

ALZHEIMER'S DISEASE

- *The most common form of dementia in the elderly is **Alzheimer's disease.** Almost 50% of individuals 85 years of age and older will suffer from this disease.*
- *Microscopically, neurons are observed to accumulate neurofibrillary tangles of filaments. Neuritic **plaques** are formed in neurons of the hippocampus, neocortex, and amygdala. These structures consist of neuritic processes surrounding a central **amyloid core.** This core consists primarily of an **Aβ peptide** derived from an **amyloid precursor protein** (APP). **Hirano bodies,** which consist of beaded filaments, accumulate in pyramidal neurons of the hippocampus of these patients.*

III. Glial Cells

A. General Features

1. **Glial cells** (glia) outnumber neurons 10 to 1 in the CNS, but because of their smaller size these cells take up less volume in the CNS than neurons (Figure 9–3).

2. Glia have the capacity to undergo **mitosis,** unlike neurons. However, populations of **progenitor cells** within the mature brain are capable of dividing and differentiating into neurons.

B. Astrocytes

1. **Astrocytes** are the largest of the glial cells (as large as 40 μm in diameter) and are the most abundant. Their nucleus is large and relatively light staining.

2. These star-shaped or stellate cells have many processes that can terminate at blood vessels and neurons.

3. **Protoplasmic astrocytes** are usually found in **gray matter** and typically cover nonsynaptic neuronal surfaces. These cells tend to have clearer cytoplasm than fibrous astrocytes.

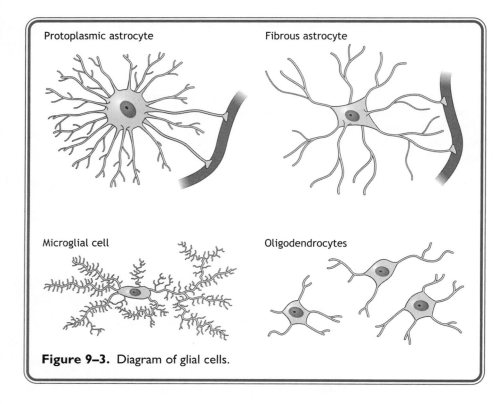

Figure 9–3. Diagram of glial cells.

4. **Fibrous astrocytes** are mostly found in **white matter.** These astrocytes have more **glial fibrillary acidic protein** (GFAP) than protoplasmic astrocytes. This protein is commonly used to detect astrocytes in brain tissue immunocytochemically.

5. Astrocytes function to repair CNS tissue or wall off a damaged area by forming a scar or **plaque.** At this site, these cells also play a role in fluid transport by their investment of blood vessels.

6. Astrocytic processes are seen in close proximity to the **node of Ranvier.** This cell may play a role in the maintenance of ion concentrations at this active region. Astrocytic processes have K^+ **channels** and Na^+, K^+-adenosine triphosphatase (ATPase).

ASTROCYTE TUMOR

- *Astrocytoma, a tumor consisting of astrocytes, is the most common form of primary brain tumors. Astrocytomas are classified based on their rate of growth.*

- *Grade I astrocytomas grow slowly, whereas grade IV astrocytomas are highly malignant.*

    C. **Oligodendrocytes**

        1. **Oligodendrocytes** are smaller than astrocytes and have a small, round, dense nucleus.

        2. **Oligodendrocytes,** found between myelinated axons, become enlarged (20 μm) during myelination and thus are found mostly in the white matter.

In the white matter of mature adults, these cells are reduced in diameter (10 μm).

3. **Satellite oligodendrocytes** are small (10 μm) and restricted to the gray matter and may play a role in neuron maintenance.

4. Oligodendrocytes in developing brain are detected immunocytochemically by the presence of myelin proteins, specifically **myelin basic protein** (MBP) and **myelin-associated glycoprotein** (MAG).

OLIGODENDROCYTE TUMOR

• *Oligodendrogliomas* *are a form of neoplasm consisting of oligodendrocytes. These neoplasms, which represent 5–10% of CNS gliomas, are usually found in the white matter of adults.*

D. Microglial Cells

1. **Microglial cells** are small, elongate cells with nuclei with condensed chromatin. These cells are found in both white and gray matter. Their nuclei are small and bean shaped.

2. Microglial cells are **phagocytic cells** within the CNS and are thought to arise from **bone marrow** stem cells.

3. These cells have short extensions and **a thorny appearance.**

E. Ependymal Cells

1. **Ependymal cells** line the ventricles of the brain and spinal cord and are interconnected by zonulae occludens.

2. These cells have apical **cilia** that function to move **cerebrospinal fluid** (CSF) within the ventricles.

3. Some ependymal cells have long processes that extend into the neural tissue called **tanycytes,** although most of these cells have a typical flattened epithelial cell shape.

4. Tanycytes are most prevalent in the floor of the **third ventricle.**

EPENDYMAL CELL TUMOR

• *Ependymomas* *are gliomas arising from cells lining the ventricles. These tumors generally arise in children younger than 5 years and represent 1–3% of intracranial neoplasms.*

IV. Nerve Fibers

A. Nerve fibers are axons that are surrounded by a specialized membrane produced by **oligodendrocytes.**

B. The membrane processes, or **myelin,** of oligodendrocytes wrap around the axons in a spiral fashion.

1. A single oligodendrocyte can myelinate as few as 3 and as many as 50 different axons, such as in the **optic nerve.**

2. The myelinated region of the axon is termed the **internode** (200–2000 nm in length), and the nonmyelinated region is called the **node of Ranvier.**

3. Adjacent to the node of Ranvier (1–2 nm wide) are terminal loops of myelin called the **paranode.**

4. Those axons not myelinated lack nodes of Ranvier but are surrounded in a random fashion by processes of neurons and astrocytes.

5. Usually large-diameter axons (20 μm in diameter) are myelinated, whereas smaller axons and dendrites are rarely myelinated.

MYELIN DISEASE

CLINICAL CORRELATION

- *Multiple sclerosis* (MS) is an *autoimmune disease* in which axons become devoid of myelin to form a plaque. The myelin is stripped from the axons by invading *macrophages.* Processes of *astrocytes* penetrate that space previously occupied by the myelin.

    C. Myelin synthesis is an adding of oligodendrocyte membrane at the leading edge of its process as it wraps around an axon at a specific site.
       1. This process then completely surrounds the axon, and its lip comes together to form a **mesaxon.**
       2. The cytoplasmic faces of each surface of the oligodendrocyte process become closely positioned to form the **major dense line** (MDL).
       3. Close apposition of the outer faces of the plasma membrane of 2 adjacent oligodendrocyte processes forms the **intraperiod line** (IPL).

    D. CNS myelin is composed of protein and lipid.
       1. In humans, about 30% of myelin is **protein,** whereas 70% is **lipid**; typical membranes have a 1:1 ratio of protein to lipid.
       2. Proteolipid protein (PLP) comprises 50% of total myelin protein and is found at the IPL of compact myelin.
       3. Myelin basic protein (MBP) comprises 30–35% of total myelin protein and is located in the MDL.
       4. MAG comprises less than 1% of myelin protein.

V. Node of Ranvier

    A. The myelinated axon is exposed to the extracellular environment at the **node of Ranvier.** This segment of the axon possesses a high density of sodium and potassium channels, and the action potential occurs at this site.

    B. An **action potential** is a short, spike-like depolarization that propagates as an electrical wave at high velocity along the axon.
       1. Action potentials occur as **sodium** ions enter the cytoplasm followed by **potassium** ions moving into the extracellular region.
       2. This wave of depolarization is conducted from 1 node of Ranvier to the next in a process called **saltatory conduction.**
       3. An **action potential** is propagated along a myelinated axon much more rapidly than an unmyelinated axon, and less energy is required to return the ion concentration to an equilibrium state.
       4. Na^+, K^+-ATPase is responsible for maintaining and returning the Na^+ and K^+ to their resting potential and is concentrated at the node of Ranvier.

VI. Gray and White Matter

    A. General Features
       1. Gray matter is composed of perikaryons, unmyelinated axons, protoplasmic astrocytes, oligodendrocytes, and microglia.
       2. White matter consists of myelinated axons, fibrous astrocytes, oligodendrocytes, and microglial cells.

B. Cerebrum

1. The **cerebrum** consists of the outer layer of gray matter, the cortex, and a core of white matter (Figure 9–4A).
2. There are about 10 billion neurons in the cerebral cortex; 1 neuron is capable of synaptic connections with as many as 100,000 other neurons (Figure 9–2B).
3. The cerebral cortex has cells that are pyramidal, stellate, or spindle shaped and arranged in **6 discrete layers.**
   a. The outermost **molecular layer** contains processes and some neurons.
   b. The **external granular layer** contains granule neurons and neuroglial cells.
   c. The **external pyramidal layer** contains pyramidal and granule neurons and neuroglial cells.
   d. The **internal granular layer** contains granule neurons and some neuroglial cells.
   e. The **internal pyramidal layer** contains pyramidal neurons and some neuroglial cells.
   f. The innermost **multiform layer** contains neuroglial cells and neurons of various shapes.
4. The cerebrum coordinates language, learning, and memorization and is responsible for integration and coordination of voluntary motor responses.

C. Cerebellum

1. The **cerebellum** consists of an outer layer of gray matter and a more central core of white matter (Figure 9–4B).
2. The 3 layers of the cerebellar cortex are the **outer molecular layer,** central **Purkinje cell layer,** and inner **granular layer.**
3. **Purkinje cells** receive both excitatory and inhibitory impulses from the motor areas of the cerebral cortex. Purkinje cells are flasklike cells that have a diameter of ~150 μm and a myelinated axon.
4. The cerebellum modulates and organizes the motor impulses to coordinate movements of muscle groups.

D. Spinal Cord

1. The **white matter** of the spinal cord is peripheral; the gray matter is central and resembles the letter H (Figure 9–4C).
2. The **central canal** is a remnant of the lumen of the embryonic neural tube and is lined by ependymal cells.
3. **Gray matter** of the ventral bars forms the anterior horns that contain motor neurons, whose axons form the ventral roots of the spinal nerves.
4. Gray matter of the posterior horns receives sensory fibers from neurons in the spinal ganglia (dorsal roots).

VII. Meninges

A. General Features

1. The brain and spinal cord are surrounded by the **meninges.**
2. The 3 layers of the meninges are the **dura mater, arachnoid,** and **pia mater.**

B. Dura Mater

1. The **dura mater** is the outer dense connective tissue adjacent to the skull. The **periosteal dura** serves as the periosteum of the inner surface of the cranial bone. This layer contains blood vessels.

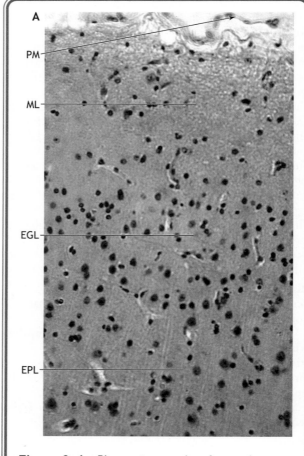

Figure 9–4. Photomicrographs of central nervous system tissue. **A:** Cerebral cortex. (continued)

2. The **meningeal dura** is located between the **periosteal dura** and the **dural border cells.**

3. Dural border cells are interconnected by a few tight junctions and immediately contact the arachnoid barrier cells.

C. Arachnoid

1. **Arachnoid barrier cells** are interconnected by tight junctions and form the **blood–CSF barrier.** The arachnoid is avascular.

2. The **trabeculae** of the arachnoid are fibroblasts that connect the arachnoid with the pia mater.

3. Cavities between the trabeculae form the **subarachnoid space,** which is filled with CSF.

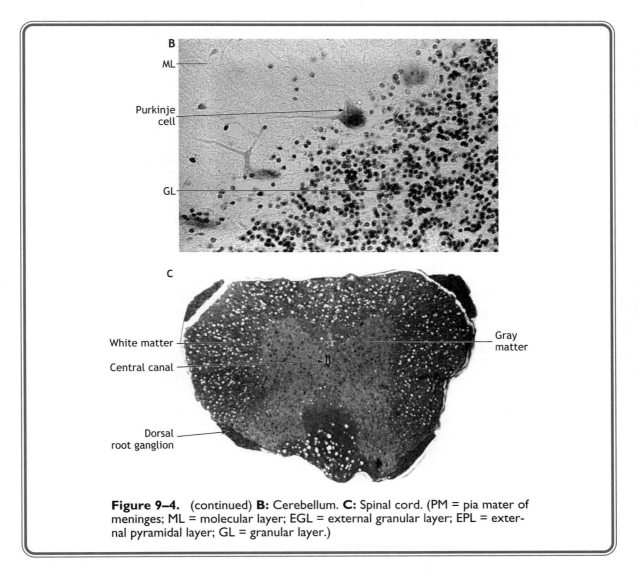

Figure 9–4. (continued) **B:** Cerebellum. **C:** Spinal cord. (PM = pia mater of meninges; ML = molecular layer; EGL = external granular layer; EPL = external pyramidal layer; GL = granular layer.)

4. Subarachnoid space communicates with the ventricles of the brain via **foramina of Luschka and Magendie.**
5. **Arachnoid villi** are fine protrusions of the arachnoid that penetrate the superior sagittal sinus. These villi function to pass CSF into venous blood.

D. Pia Mater
  1. The **pia mater** consists of flattened cells with loose connective tissue and blood vessels.
  2. This layer invests the brain and is the only meningeal layer that extends into sulci of the brain.

3. Between the pia mater and neural elements is a thin layer of protoplasmic as-trocytic processes that firmly adheres to the pia mater, referred to as the **glial limitans.**

MENINGEAL TUMORS AND DISEASE

- *Meningiomas* are slow-growing tumors of cells of the **chorionic villi.** These tumors are usually lo-cated adjacent to the dura mater. This tumor may increase the intracranial pressure.
- *Meningitis* is an inflammation of the meninges. This condition arises from an infestation of bacteria or viruses. **Viral meningitis** is caused by the mumps virus and coxsackievirus, whereas **bacterial menin-gitis** is caused by the influenza and pneumococcal bacteria. The choroid plexus ultimately becomes congested with white blood cells, and endothelial cells swell, which restricts CSF flow.

VIII. Choroid Plexus and Cerebrospinal Fluid

A. Choroid Plexus

1. The **choroid plexuses** are invaginated folds of pia mater that penetrate the interior of brain third and fourth ventricles and lateral ventricles.
2. The choroid plexus consists of loose connective tissue covered by a simple cuboidal or low columnar epithelium that is continuous with the **ependyma.** The choroid is highly vascularized and consists of dilated, fenestrated capil-laries.

B. Cerebrospinal Fluid

1. **CSF** is secreted by epithelial cells (70%) that line the choroid plexus. It is ac-tually an ultrafiltrate of plasma that is modified within the epithelial cells and then secreted into the ventricles.
2. CSF is found in the **subarachnoid space, ventricles** of the brain, and **cen-tral canal** of the spinal cord. Within these systems, CSF amounts to about 125–150 mL; however, 400–500 mL is produced daily.
3. This fluid is clear and colorless with little protein content and 2–5 lympho-cytes/mL.
4. The **functions** of CSF include protection and support by its buoyancy, maintenance of normal homeostasis, elimination of metabolic wastes, and transport within the CNS.

CEREBROSPINAL FLUID DISORDER

- *Blockage* of the movement of CSF out of the ventricles through foramina or into the venous circula-tion via *arachnoid villi* results in the condition called **hydrocephalus.** This condition leads to ex-panded ventricles, resulting in damage to brain tissue or expansion of the skull in infants.

CLINICAL PROBLEMS

Your patient received blunt head trauma that resulted in memory loss and intracranial swelling. You suspect that cells of the brain were extensively damaged.

1. Which of the following cells phagocytize the cellular debris created by neural damage?

   A. Fibrous astrocytes

   B. Oligodendrocytes

   C. Protoplasmic astrocytes

   D. Ependymal cells

   E. Microglial cells

2. Which of the following form the choroid plexus?

   A. Pia mater and arachnoid

   B. Arachnoid cells and venous endothelium

   C. Pia mater and the overlying ependyma

   D. Arachnoid barrier cells

   E. Meningeal dura

3. Which of the following serve to allow CSF to exit the subarachnoid space?

   A. Arachnoid villi

   B. Choroid plexus

   C. Arachnoid barrier cells

   D. Vessels within the subarachnoid space

   E. Fenestrated capillaries

A pathologist is examining a stained section of an unknown nervous tissue. He notices cells with an extremely large, flask-shaped cell body and multiple dendrites lying in parallel.

4. Which of the following tissues is the pathologist examining?

   A. White matter of the spinal cord

   B. Gray matter of the spinal cord

   C. Cerebral cortex

   D. Cerebellar cortex

   E. White matter of cerebrum

5. During early brain development, myelination of axons is a critical event. Which of the following cells is responsible for this important function?

   A. Fibrous astrocytes

   B. Protoplasmic astrocytes

   C. Oligodendrocytes

   D. Schwann cells

   E. Microglial cells

6. During development of the CNS, which of the following would result from the synthesis of a nonfunctional or lack of myelin basic protein (MBP)?

A. No intraperiod line

B. No major dense line

C. No mesaxon

D. Longer than normal nodes of Ranvier

E. Increased velocity of action potentials

A brain tumor of unknown origin was surgically removed from an elderly male. Radiation therapy was administered to ensure that tumor cells not removed do not repopulate the brain.

7. Which of the following cells would be decreased in number after the radiation therapy?

A. Oligodendrocytes

B. Protoplasmic astrocytes

C. Microglial cells

D. Neurons

E. Fibrous astrocytes

8. Which of the following events occurs immediately after an action potential reaches an axon terminal at its synapse?

A. Vesicle fusion with the presynaptic terminal membrane

B. Calcium ion influx at the presynaptic terminal

C. Neurotransmitter binding to the receptor on the postsynaptic terminal

D. Neurotransmitter release into the synaptic cleft

E. Binding of the neurotransmitter to the presynaptic terminal

A 60-year-old woman has suffered from a neurologic disorder for several years. Analysis of the CSF revealed the presence of antibodies to myelin proteins.

9. From which of the following disorders does she suffer?

A. Multiple sclerosis

B. Amyotrophic lateral sclerosis

C. Parkinson's disease

D. Guillain-Barré syndrome

E. Oligodendroglioma

A brain tumor is removed and analyzed by immunocytochemistry. It was shown that the tumor expressed high levels of glial fibrillary acidic protein (GFAP).

10. Which of the following tumors was being analyzed?

A. Oligodendroglioma

B. Astrocytoma

C. Ependymoma

D. Neuroblastoma

E. Meningioma

Tissue from the CNS of an 80-year-old man is examined by electron microcopy. Cells in this tissue sample have several residual bodies and lipofuscin within their cytoplasm.

11. Which of the following cells are being described?

    A. Endothelial cells of capillaries

    B. Astrocytes

    C. Neurons

    D. Oligodendrocytes

    E. Fibroblasts of dura mater

12. Which of the following cellular structures is found within the cytoplasm of perikarya but is not within axoplasm?

    A. Membrane-bound vesicles

    B. Neurofilaments

    C. Microtubules

    D. Rough endoplasmic reticulum

    E. Mitochondria

ANSWERS

1. The answer is E. Microglial cells phagocytize cellular debris and dead cells. These cells are thought to originate from blood cell precursors. Astrocytes may have limited phagocytic property but are primarily a supportive cell through release of factors and process formation. Oligodendrocytes primarily function in myelin formation, whereas ependymal cells line the ventricles and central canal of the spinal cord.

2. The answer is C. Choroid plexuses are formed by the extension of the pia mater and choroidal vessels into the ependyma. The dura mater and the arachnoid are not components of choroid plexuses.

3. The answer is A. CSF courses from the ventricles into the subarachnoid space. From the subarachnoid space, CSF passes through arachnoid villi into the superior sagittal sinus.

4. The answer is D. Large flask-shaped cells within nervous tissue are Purkinje neurons. These cells have an extensive dendritic tree and are found within the cerebellar cortex between the molecular and granular layers.

5. The answer is C. Oligodendrocytes are the myelin-forming cells within the CNS. Schwann cells myelinate axons within the peripheral nervous system.

6. The answer is B. In the CNS, myelin basic protein (MBP) is responsible for the formation major dense line (MDL), which forms the close apposition of inner cytoplasmic regions of oligodendrocyte membranes. Without the MDL, myelin formation is reduced to oligodendrocytes wrapping the axon and no nodes of Ranvier, resulting in a decrease in the velocity of the action potential.

7. The answer is D. After radiation therapy, the only cells that would repopulate the brain must have a mitotic potential such as astrocytes, some oligodendrocytes, pericytes, and microglial cells. Neurons are postmitotic cells and thus would not regenerate within the brain.

8. The answer is B. Immediately after the action potential reaches the axon terminus, calcium channels open to allow for the influx of calcium ions. The other events occur after calcium ion influx. Neurotransmitters bind to receptors on the postsynaptic membrane, not the presynaptic membrane.

9. The answer is A. Multiple sclerosis is an autoimmune disorder marked by an attack of macrophages on myelin. The CSF of MS patients contains antibodies to myelin proteins.

10. The answer is B. The immunocytochemical marker for astrocytes is glial fibrillary acidic protein (GFAP). No other cell within the brain normally expresses this protein.

11. The answer is C. Neurons in some regions of an aged brain will accumulate residual bodies and lipofuscin. Both of these cytoplasmic inclusions arise because of intracellular metabolism and autophagic activity.

12. The answer is D. The axoplasm contains most of the organelles found within the cytoplasm of the perikaryon with the exception of rough endoplasmic reticulum.

CHAPTER 10
PERIPHERAL NERVOUS SYSTEM

I. Peripheral Nerves

A. General Features

1. **Peripheral nerve fibers** are arranged in bundles, or fascicles, called **nerves** (Figure 10–1A).

2. The **epineurium** is the external fibrous coat that surrounds the peripheral nerve and also fills the space between nerve fiber bundles. It consists of collagen, fibroblasts, and blood vessels.

3. The **epineurium** provides both structural support and elasticity.

4. Each nerve bundle is surrounded by a **perineurium,** which consists of flattened epithelial-like cells. The epithelial cells are joined by tight junctions that protect the nerve fiber from toxic macromolecules.

5. **Schwann cells** wrap around each nerve fiber, which is surrounded by a connective tissue layer containing reticular cells, called the **endoneurium.**

B. Myelination

1. **Schwann cells** myelinate nerve fibers of the peripheral nervous system (PNS) at a single site along an axon.

2. **Nodes of Ranvier** lie between consecutive Schwann cells.

3. Some areas of myelin do not compact, leaving small areas of Schwann cell cytoplasm, referred to as **Schmidt-Lanterman clefts** (Figure 10–1B).

4. Early in the myelination process, Schwann cells enclose many axons in a trough indenting its surface. After repeated mitotic events, Schwann cells become segregated to a single axon.

5. The lips of the Schwann cell process extend around the axon in a spiral fashion forming the **intraperiod line.**

6. As the cytoplasm begins to be lost during the spiraling process, the **major dense line** is formed.

7. As the sheath becomes more mature, the number of myelin turns increases and the lamellae of the sheath become more **compact.**

C. Components of PNS Myelin

1. Human PNS myelin consists of **30% protein** and **70% lipid.**

2. The major protein in PNS myelin is the **Po protein** (50–60%), which is transmembrane protein found at the MDL and IPL.

3. **Myelin basic proteins** make up 20% of human PNS myelin and are found within the MDL.

4. Myelin basic protein (MBP) is not essential for formation of the **MDL** in PNS myelin because of the transmembrane Po protein.

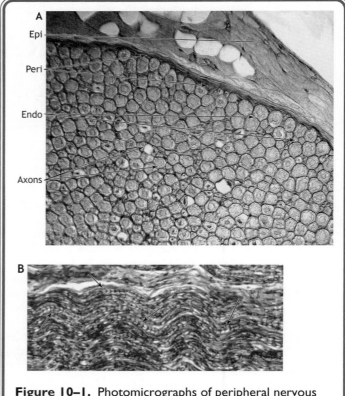

Figure 10–1. Photomicrographs of peripheral nervous system tissue. **A:** Peripheral nerve (Epi = epineurium; Peri = perineurium; Endo = endoneurium). **B:** Schmidt-Lanterman clefts (arrows). (continued)

GUILLAIN-BARRÉ SYNDROME

· *Guillain-Barré syndrome is a disorder marked by **demyelination** and inflammation of peripheral nerves and motor fibers of ventral roots. Muscle weakness in the extremities is a symptom of this disorder.*

 II. **Ganglia**

 A. A collection of nerve cell bodies outside the central nervous system (CNS) is called a **ganglion.**

 B. Ganglia are surrounded by connective tissue, and each neuron is surrounded by a **satellite cell,** which functions as a support cell.

 C. The satellite cell is surrounded by a basal lamina.

 D. There are 2 types of nerve ganglia based on morphology and function. The direction of the impulse determines whether the ganglion is a **sensory** or an **autonomic** ganglion.

   1. **Sensory ganglia** are cranial when associated with cranial nerves or spinal when associated with the dorsal root ganglia of spinal nerves (Figure 10–1C).

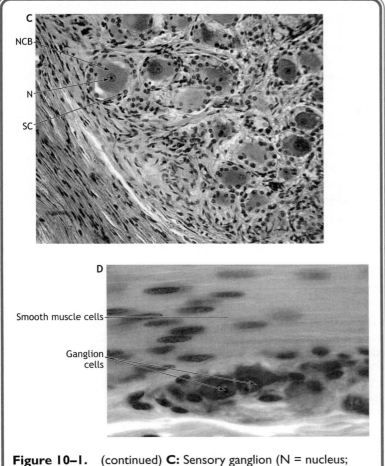

Figure 10–1. (continued) **C:** Sensory ganglion (N = nucleus; SC = satellite cell; NCB = neuron cell body). **D:** Ganglion cells of myenteric plexus of Auerbach.

    a. Dorsal root ganglia have **pseudounipolar neurons** that send 1 process to the CNS and the other to the periphery. These neurons have large (100 μm) cell bodies.

    b. The perikarya of the ganglia do not receive impulses; thus, **no synapses** are found in these ganglia, and their nuclei are centrally located.

    c. Nerve fibers are found in **bundles,** and nerve cell bodies are **grouped in clusters** at the periphery of these ganglia.

2. Autonomic ganglia are found in the sympathetic and parasympathetic divisions of the autonomic nervous system.

    a. These ganglia are found in the walls of the digestive tract, called the **myenteric plexus of Auerbach,** and the **submucosal plexus of Meissner,** called intramural ganglia (Figure 10–1D).

b. The **multipolar neurons** of autonomic ganglia are surrounded by a layer of satellite cells, but in intramural ganglia only a few satellite cells are found.

    c. The nucleus of autonomic neurons is round and eccentrically located, and the cytoplasm contains fine Nissl bodies.

    d. Cell bodies of neurons in autonomic ganglia are not grouped but are found scattered between nerve fibers, which show synaptic connects.

III. Autonomic Nervous System

A. The autonomic nervous system (ANS) controls smooth muscle contraction, secretion by some glands, and cardiac rhythm.

B. The ANS consists of nerves within the CNS, fibers that leave the CNS through cranial and spinal nerves, and nerve ganglia in the paths of these fibers.

C. The first neuron of the ANS is located in the CNS and is called **preganglionic;** the axon of the second neuron to the effector organ is called **postganglionic.**

D. **Norepinephrine** is the chemical transmitter at most sympathetic postganglionic endings.

E. **Acetylcholine** is released by the preganglionic and postganglionic nerve endings of the parasympathetic system.

IV. Degeneration and Regeneration of Neurons

A. **Degeneration**

    1. Neurons are not able to divide; thus, their degeneration is a permanent loss, although some cells within the PNS are able to divide, including **Schwann cells.**

    2. Degenerative changes occur, called **Wallerian degeneration,** after damage or severing of a peripheral nerve axon.

    3. If the **proximal** segment maintains its continuity with the perikaryon, it often regenerates but initially degenerates 1–2 internodes.

    4. The distal segment degenerates and is removed by Schwann cells or macrophages.

    5. Axonal injury causes dissolution of the Nissl substance, called **chromatolysis,** which results in a decrease in cytoplasmic basophilia, an increase in perikaryon volume, and migration of the nucleus to a peripheral position in the perikaryon.

B. **Regeneration**

    1. Most essential for neuron regeneration, **Schwann cells proliferate,** giving rise to tubelike structures that are surrounded by collagen of the **endoneurium.**

    2. Each axon from the surviving neuron cell body splits into fibers and reaches the Schwann cell tube of the degenerating stump. These tubes serve as a guide for axonal growth.

    3. Some axons reach peripheral tubes, whereas other axons reach central tubes and become completely surrounded by plasma membrane of Schwann cells.

    4. Only 1 fiber ultimately persists and becomes myelinated, which is usually the **largest fiber.**

    5. The enlargement of 1 fiber and the elimination of others occur only if the regenerating axons make sensory or motor contact with the appropriate **target.**

AMYOTROPHIC LATERAL SCLEROSIS

• *Amyotrophic lateral sclerosis,* also called **Lou Gehrig disease,** is a nerve disorder marked by degeneration of upper and lower motor neurons. This disease is usually fatal within 2–3 years.

V. Nerve Endings and Organs of Special Sense

A. Nonencapsulated Receptors

    1. Free nerve endings

        a. Free nerve endings are small fibers that may or may not be myelinated, but their terminations are devoid of myelin sheaths and Schwann cells (Figure 10–2).

        b. These endings, usually within **epidermis** of skin, are responsible for pain and touch sensations.

    2. Merkel's disk

        a. Merkel's disk is a specialized epithelial cell mass located in the **basal layer** of the epidermis, which is enmeshed in a terminal branch of underlying axon terminals.

        b. Each mass contacts a **tactile cell** and is sensitive to **touch.**

    3. Nerve endings of hair follicles

        a. Unmyelinated fibers surround the outer connective tissue sheath of the **hair follicle.**

        b. The sensation of touch is affected when the hair shaft is perturbed or bent.

B. Encapsulated Receptors

    1. Meissner's corpuscle

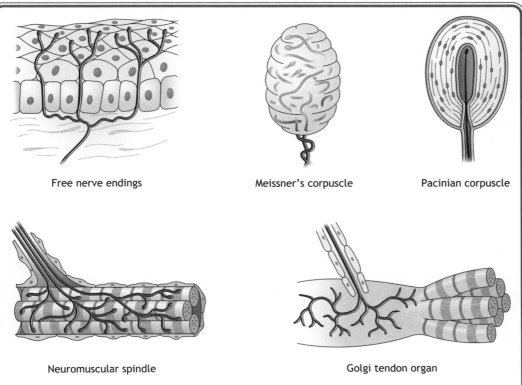

Free nerve endings   Meissner's corpuscle   Pacinian corpuscle

Neuromuscular spindle   Golgi tendon organ

Figure 10–2. Diagrams of organs of special sense.

a. Meissner's corpuscles are beehive-shaped structures found just below the epidermis of the skin in the **dermal papillae.**

    b. These structures are a stack of flattened cells enclosed by a capsule and sensitive to touch.

2. Pacinian corpuscle

    a. Pacinian corpuscles are widely distributed throughout the body and respond to pressure, tension, and possibly vibrations.

    b. This corpuscle, found within the **dermis,** is a cylindrical core surrounded by 20–70 concentric lamellae of fibroblasts and thin collagen fibers.

    c. Its axon passes through the corpuscle center and terminates in an expanded end.

3. Neuromuscular spindle

    a. Neuromuscular spindles are found in skeletal muscle and function to relay sensory information to the CNS for controlling muscle activity to maintain posture or regulate the activity of opposing muscle groups (muscle tone) involved in motor activities, such as walking.

    b. These spindles, 0.1 mm wide and 1–5 mm long, consist of a fluid-filled space with modified muscle fibers encased in groups within a connective tissue shell.

    c. Muscle fibers lying outside the spindle are called **extrafusal** fibers, whereas those fibers inside the spindle capsule are called **intrafusal fibers** (3–5 per muscle fiber).

    d. Intrafusal fibers are **nuclear chain fibers,** which are short, small-diameter fibers, whereas **nuclear bag fibers** are larger fibers.

    e. Intrafusal fibers are richly innervated by sensory axons that monitor the degree of **stretch** placed on these fibers.

4. Golgi tendon organs

    a. Golgi tendon organs are cylindrical structures found within the tendon at its junction with muscle fibers.

    b. These structures monitor the **force** of a muscle contraction and provide an inhibitory feedback on the contraction of this muscle.

    c. The Golgi tendon organs and neuromuscular spindles work in concert to integrate **spinal cord reflex systems.**

CLINICAL PROBLEMS

1. Which of the following structures prevent toxic materials from attacking an axon of a peripheral nerve?

    A. Perineurium

    B. Endoneurium

    C. Clefts of Schwann cells

    D. Epineurium

    E. Neurolemma

2. Which of the following best characterizes Schmidt-Lanterman clefts?

   A. Site nearest node of Ranvier

   B. Schwann cell cytoplasm

   C. Region between nodes of Ranvier

   D. Axoplasm

   E. Region lacking myelin

3. Which of the following is a primary component or characteristic of a sensory ganglion?

   A. Scattered neuron cell bodies

   B. Multipolar neurons

   C. Synaptic contacts

   D. Neurons with eccentric nuclei

   E. Pseudounipolar neurons

4. Which of the following receptors consists of concentric connective tissue with a central axon?

   A. Merkel's disk

   B. Neuromuscular spindle

   C. Meissner's corpuscle

   D. Pacinian corpuscle

   E. Nerves surrounding a hair follicle

5. Which of the following is the most critical event in regeneration of a peripheral nerve?

   A. Hypertrophy of Schwann cells

   B. Chromatolysis of the damaged perikaryon

   C. Proliferation of Schwann cells

   D. Phagocytic activity of macrophages

   E. Distal axon degeneration

6. Merkel's disks are found in which of the following regions?

   A. Between smooth muscle layers in the intestine

   B. Stratum basale of skin

   C. Dermis of skin

   D. Submucosa of the intestine

   E. Hypodermis

7. Which of the following components of the meninges are most similar to the endoneurium that covers a peripheral nerve?

   A. Pia mater

   B. Arachnoid border cells

   C. Subarachnoid space

D. Meningeal dura

   E. Dural border cells

8. Which of the following best characterizes the difference in myelinated axons of the PNS compared with the CNS?

   A. No intraperiod line

   B. No myelin basic protein

   C. All axons surrounded by Schwann cells

   D. Schwann cells myelinate at multiple axonal sites

   E. All axons have nodes of Ranvier

9. Which of the following structures primarily functions to monitor the degree of stretch of muscles?

   A. Merkel's disk

   B. Golgi tendon organ

   C. Myoneural junction

   D. Pacinian corpuscle

   E. Neuromuscular spindle

10. Which of the following cells of the PNS functions equivalent to the oligodendrocyte within the CNS?

   A. Fibroblasts

   B. Schwann cells

   C. Satellite cells

   D. Macrophages

   E. Endothelial cells

ANSWERS

1. The answer is A. The perineurium consists of cells that are joined by tight junctions, thereby preventing toxic substances from attacking a nerve fiber.

2. The answer is B. Schmidt-Lanterman clefts are sites in which myelin is not compacted; thus, the major dense line is not formed. These areas allow for substances synthesized within the cell body of the Schwann cell to migrate to those regions of myelin requiring cell membrane replacement.

3. The answer is E. Sensory ganglia contain pseudounipolar neurons, but autonomic ganglia have multipolar neurons. No synapses are found in sensory ganglia, and the neurons are organized in clumps.

4. The answer is D. Receptors that are organized as a concentric arrangement of connective tissue are Pacinian corpuscles. Meissner's corpuscles are shaped like a beehive and have a large cellular component.

5. The answer is C. After peripheral nerve damage, the most critical event that must occur that allows axons to regenerate is the proliferation of Schwann cells. These cells then create tunnels through which sprouts of these axons can travel to ultimately interact with their target. All the other events precede Schwann cell proliferation or are a consequence of axonal damage.

6. The answer is B. Merkel's disks are found in the stratum basale, also called stratum germinativum, of the epidermis.

7. The answer is A. The epineurium is the immediate layer of a peripheral nerve. The pia mater is the layer of the meninges in immediate contact with the brain and follows all convolutions of the brain.

8. The answer is C. All axons within the PNS are surrounded by processes of Schwann cells; however, not all axons are myelinated or have nodes of Ranvier.

9. The answer is E. The neuromuscular spindle consists of intrafusal fibers that monitor the degree of stretch on a muscle. In contrast, the Golgi tendon organ functions to monitor the force generated by muscle contraction.

10. The answer is B. The Schwann cell of the PNS myelinates axons, as also performed by oligodendrocytes with the CNS. The fibroblast is a component of the PNS but synthesizes collagen and other extracellular proteins. Satellite cells surround neuron cell bodies and are protective. Macrophages are phagocytic cells that also secrete growth factors, and endothelial cells line the lumen of blood vessels.

CHAPTER 11
CARDIOVASCULAR SYSTEM

I. Components of Cardiovascular System

A. The heart and the great vessels of the heart and other large vessels constitute the **gross vascular system.**

B. Capillaries, arterioles, and venules make up the **microvascular system.**

II. Capillaries

A. Capillaries are **low-pressure vessels** that permit passive diffusion across their walls and consist of a single layer of endothelium and possibly a basal lamina and pericytes (Figure 11–1A).

B. Capillaries are divided into 3 types according to the **integrity of their endothelial cell lining.**

    1. Continuous capillaries are formed by single layers of endothelium rolled up to form a tube. Tight junctions and a basal lamina are present. Endothelial cells contain pinocytotic vesicles. These capillaries are found in muscle, connective tissue, central nervous system, and gonads.

    2. Fenestrated capillaries have pores or fenestrae in the endothelial cell wall, sometimes covered by a thin diaphragm. These capillaries are found in the gastrointestinal tract, endocrine glands, and renal glomeruli.

    3. Discontinuous capillaries (also called **sinusoidal capillaries**) have a discontinuity or gap in the endothelial cell lining; thus, the basal lamina is incomplete or missing. These capillaries are found in spleen, bone marrow, and liver.

C. The **endothelium** of capillaries functions in the exchange of gases, enzyme-mediated reactions, exchange of fluids and metabolites via small and large pores, and phagocytosis.

D. Pericytes surround capillaries at irregular intervals. These cells are thought to have **contractile** properties because of the presence of tropomyosin and myosin within their cytoplasm.

III. Layers of Vessels

A. Tunica Intima

    1. The **tunica intima** consists of an endothelium, a simple squamous epithelium, which faces the lumen of a vessel (Figure 11–2).

    2. A **subendothelial** connective tissue and an internal elastic lamina may or may not be present.

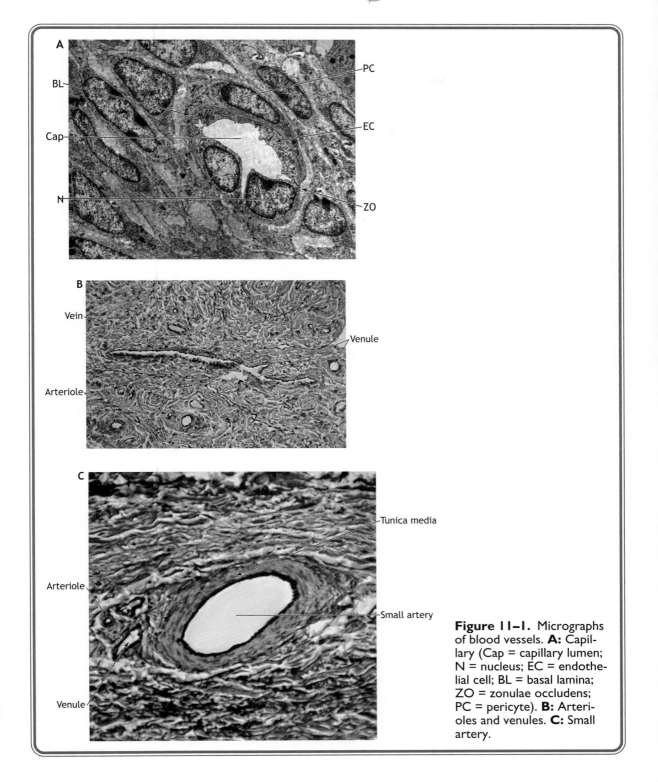

Figure 11–1. Micrographs of blood vessels. **A:** Capillary (Cap = capillary lumen; N = nucleus; EC = endothelial cell; BL = basal lamina; ZO = zonulae occludens; PC = pericyte). **B:** Arterioles and venules. **C:** Small artery.

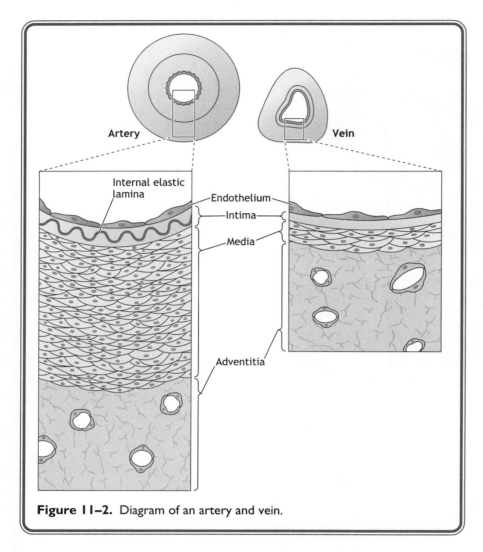

Figure 11–2. Diagram of an artery and vein.

B. Tunica Media
1. Smooth muscle cells of the **tunica media** are oriented circumferentially around a vessel (Figure 11–2).
2. **Elastic fibers** synthesized by smooth muscle cells are present in large arteries, such as the aorta.
3. An **external elastic lamina,** which is not considered part of either tunic, separates the tunica media from the tunica adventitia.
4. A **vasa vasorum** consists of blood vessels that nourish the tunica media.

C. Tunica Adventitia
1. The **tunica adventitia** has protective and nutrient functions and consists primarily of connective tissue (Figure 11–2).
2. A **vasa vasorum** consists of blood vessels that nourish the tunica adventitia.

IV. Comparison of Companion Arteries and Veins

A. The wall of a vein is usually thinner than its companion artery.

B. The lumen of a vein is often irregular or collapsed, and larger veins have valves.

C. Tunica adventitia is predominant in large veins, whereas the tunica media is predominant in large arteries.

V. Classification of Arteries

A. Arterioles

1. **Arterioles** lack a subendothelial layer and an internal elastic limiting membrane (Figure 11–1B and C).
2. One or 2 layers of **smooth muscle** cells are found in the tunica media.
3. The **tunica adventitia** is limited or absent and lacks an external elastic limiting membrane.
4. Arterioles function as control valves on capillary beds under autonomic nervous control.

B. Small Arteries

1. **Small arteries** have a tunica media of more than 2 layers of smooth muscle cells and an internal elastic limiting membrane (Figure 11–1C).
2. These arteries assist in the control of blood pressure.

C. Medium Arteries

1. The **tunica media** of medium (or **muscular**) arteries has up to 40 smooth muscle layers.
2. An internal and external elastic limiting membrane is present.

D. Large Arteries

1. The **tunica intima** of large (**elastic**) arteries is composed of endothelium, a thin connective tissue layer, and a prominent internal elastic limiting membrane.
2. A thick tunica media has abundant elastic fibers arranged in fenestrated laminae.
3. The **tunica adventitia** is composed of elastic connective tissue and is nourished by the **vasa vasorum.**
4. These arteries affect diastolic blood pressure.

VI. Classification of Veins

A. Venules

1. **Venules** have an endothelial lining, no tunica media, and a small collagenous tunica adventitia (Figure 11–1B).
2. **Histamine** causes separation of endothelial cells of venules to expose the basal lamina that allows passage of cells and fluid.
3. **Neutrophils** attach and cross the basement membrane to enter the connective tissue via **diapedesis.**

B. Small Veins

1. **Small veins** are similar to venules (Figure 11–1B).
2. Small veins have a thicker tunica media and tunica adventitia than venules.

C. Medium Veins

1. **Medium veins** have circularly arranged smooth muscle in their tunica media and longitudinally arranged smooth muscle in the tunica adventitia (Figure 11–1B).

2. The tunica media of these veins is much thicker relative to the size of the lumen than is the media of a large vein.

3. **Valves** of veins consist of 2 outpocketings of the endothelium of the **tunica intima** that extend into the lumen.

   D. Large Veins

      1. The **tunica adventitia of large veins** is most prominent and contains numerous longitudinally arranged bundles of smooth muscle (Figure 11–1B).

      2. Circularly arranged smooth muscle fibers are found in a narrow tunica media.

      3. A **vasa vasorum** is found in the tunica adventitia, and valves are generally present.

ATHEROSCLEROSIS

- *Atherosclerosis* is a disease characterized by a degenerating metabolic lesion of large arteries in which the **tunica intima** becomes thickened by cholesterol deposits and hardened by healing and degeneration.

- If **coronary arteries** are involved, circulation to the heart muscle is reduced, resulting in compromise of normal function because of *ischemia* (inadequate blood supply). The endothelium and smooth muscle are primarily involved. In early stages the effects are partly reversible, whereas at late stages vessels become irreversibly calcified and hardened.

VII. Heart

   A. Endocardium

      1. The **endocardium** of the heart is comparable to the tunica intima of arteries (Figure 11–3).

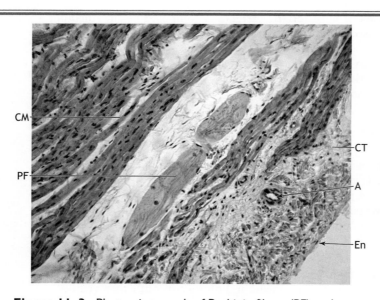

Figure 11–3. Photomicrograph of Purkinje fibers (PF) and cardiac muscle (CM = myocardium). (A = arteriole; CT = connective tissue; En = endocardium.)

2. The endocardium is composed of an endothelial lining, basal lamina, and subendocardial layer of connective tissue. Veins, nerves, and the impulse-conducting system are present in the subendocardial region
3. **Cusps** of heart valves are composed of an endocardium.

B. **Myocardium**
 1. The **myocardium** is comparable to a tunica media.
 2. It is composed exclusively of **cardiac muscle fibers.**

C. **Epicardium**
 1. The **epicardium** is also comparable to a tunica adventitia.
 2. It has a serous covering of mesothelium and a loose connective tissue subepicardial layer.

D. **Atrioventricular and Sinoatrial Nodes**
 1. **Atrioventricular (AV)** and **sinoatrial (SA) nodes** are part of the impulse-conducting system of the heart.
 2. These nodes are composed of specialized aggregates of cardiac muscle cells.
 3. The SA node is in the wall of the superior vena cava and activates the atrial myocardium.
 4. The AV node triggers impulses to the ventricles by specialized nodal cells. The AV node transmits an impulse by modified cardiac muscle cells called **Purkinje fibers** (Figure 11–3).

CARDIOVASCULAR DISEASE

- *Cardiovasular disease* is the leading cause of death in the United States and is not limited to the elderly. Cardiovascular disease afflicts more than 28 million people resulting in 1 million deaths per year.
- Damage to the AV or SA node results in **bundle branch block.** This condition is characterized by a failed excitation to affected branch, which then causes asynchronous and irregular ventricular contraction.
- Damage to the AV node may result in **heart block,** which is characterized by ventricles that beat at their own dissociated, slow rate.

TUMORS OF BLOOD VESSELS

- *Angiosarcomas* are rare tumors originating from endothelial cells.
- *Hemangiopericytomas* are usually benign and probably originate from pericytes.

CLINICAL PROBLEMS

1. Which of the following are described as low-pressure vessels that contain pinocytotic vesicles in the cytoplasm of its endothelial cells?
 A. Venules
 B. Arterioles
 C. Continuous capillaries
 D. Fenestrated capillaries
 E. Discontinuous capillaries

2. Which of the following is found within the tunica media of largest arteries?

   A. Vasa vasorum

   B. Elastic fibers

   C. Simple squamous epithelium

   D. Mast cells

   E. Fibroblasts

3. Which of the following vessels possesses a vasa vasorum in its tunica adventitia?

   A. Venule

   B. Arteriole

   C. Small artery

   D. Small vein

   E. Elastic artery

4. Which of the following are components that form valves in medium and large veins?

   A. Tunica intima only

   B. Tunica intima and media

   C. Tunica media and adventitia

   D. Tunica media only

   E. External and internal elastic membranes

5. The impulse-conducting system within cardiac muscle is composed of which of the following?

   A. Cells of the endothelium

   B. Myelinated axons

   C. Unmyelinated axons

   D. Purkinje fibers

   E. Sarcoplasmic reticulum

While examining a tissue with a light microscope, you notice a small vessel that has 2 circular layers of smooth muscle in its tunica media. The lumen of this vessel is only slightly larger than the diameter of an erythrocyte.

6. Which of the following types of vessels are you examining?

   A. Venule

   B. Fenestrated capillary

   C. Arteriole

   D. Muscular artery

   E. Large vein

7. Which of the following cells surround blood capillaries and have a contractile function?

A. Fibroblasts

B. Pericytes

C. Smooth muscle fibers

D. Purkinje fiber

E. Macrophage

8. The aorta is a prominent blood vessel within the body. Which of the following best categorizes this vessel?

A. Elastic artery

B. Muscular artery

C. Large vein

D. Medium artery

E. Medium vein

9. Sinusoids within such organs as the spleen and bone marrow are classified as which of the following types of vessels?

A. Lymph vessels

B. Veins

C. Capillaries

D. Arterioles

E. Vasa vasorum

While examining a tissue with a light microscope, you note a vessel that has no smooth muscle. However, you note a large amount of connective tissue at the periphery of the vessel.

10. Which of the following vessels are you examining?

A. Arteriole

B. Venule

C. Elastic artery

D. Capillary

E. Large vein

ANSWERS

1. The answer is C. Continuous capillaries are low-pressure vessels that have endothelial cells joined by tight junctions. Thus, substances have to pass through the endothelium via pinocytotic vesicles to reach the lumen or interstitial tissue.

2. The answer is B. Elastic fibers are found in the tunica media of large arteries; thus, these vessels are called elastic arteries, which include the aorta and its immediate branches.

3. The answer is E. Vasa vasorum is found in the tunica adventitia of large vessels, including large arteries (aorta) and veins (superior and inferior venae cavae).

4. The answer is A. Valves in veins are formed by folds of the tunica intima that extend into the lumen.

5. The answer is D. Electrical impulses are conducted within the myocardium by modified cardiac muscle fibers called Purkinje fibers. Cardiac muscle does not have myoneural junctions.

6. The answer is C. Arterioles are small arteries that have 1–2 layers of smooth muscle in the tunica media.

7. The answer is B. Pericytes and their processes surround capillaries. Because of the presence of contractile proteins within their cytoplasm, these cells are thought to play a contraction role.

8. The answer is A. The tunica media of the aorta contains large numbers of elastic fibers and thus is termed an elastic artery.

9. The answer is C. Sinusoids are a type of discontinuous capillary that allows cells to freely pass into the blood stream. These vessels are found within the bone marrow and spleen.

10. The answer is B. A venule has no smooth muscle, and thus no distinctive tunica media, but it does have a delicate tunica adventitia of connective tissue at its periphery.

CHAPTER 12
IMMUNE SYSTEM

I. General Features

A. The **immune system** protects the body against pathogenic organisms and aberrant cells.

B. The epithelial lining of organs provides the first line of defense against invading pathogens, whereas the immune system provides the second and third lines of defense.

C. Lymphoid tissue consists of individual **lymphocytes,** diffuse or densely packed masses as in **nodes** or lymphoid **organs** such as the thymus and spleen.

II. Diffuse Lymphoid Tissue

A. **Diffuse lymphoid tissue** is not sharply delineated from its surrounding connective tissue and thus has no special organization.

B. This tissue is found beneath the epithelium of the digestive tract (gut-associated lymphoid tissue, or **GALT**), respiratory tract (bronchial-associated lymphoid tissue or **BALT**), and general mucosa (mucosal-associated lymphoid tissue**, or MALT**).

C. The primary components of this tissue include **reticular stroma** and **free cells** such as macrophages, lymphocytes, and plasma cells.

III. Lymph Follicles

A. Nodular Lymphoid Tissue
   1. **Nodular lymphoid tissue** consists of dense aggregations of lymphoid tissue.
   2. These dense aggregations are arranged in spherical masses surrounded by diffuse lymphoid tissue.

B. Primary Follicle
   1. A **primary follicle** consists of mitotically active, immature B cells (**lymphoblasts**) that have not been exposed to antigens, but it lacks a germinal center.
   2. Other components of this follicle are supportive reticular cells, dendritic cells, and macrophages.

C. Secondary Follicle
   1. A **secondary follicle** contains a light-stained **germinal center** of B lymphoblasts, plasma cells, and reticular cells that forms in response to an antigen.

2. The dark peripheral region, or **corona,** consists of tightly packed B lymphocytes.
3. Aggregations of follicles are found within the palatine, pharyngeal, and lingual **tonsils, Peyer's patches** in the small intestine, and **appendix.**

IV. Lymph Node

A. General Features

1. **Lymph nodes** are ovoid filters of lymph with a distinct **cortex** and **medulla** and lymph **sinuses** (Figure 12–1A).
2. Lymph nodes are surrounded by a dense, collagenous **capsule** that sends **trabeculae** into its parenchyma.
3. The **hilus** of a lymph node is the area where vessels enter and leave the node.
4. Lymphocytes and other free cells are held in a reticular meshwork of **reticular fibers** and macrophages.

B. Cortex

1. The **cortex** is the outer, dense mass of lymphoid cells containing some trabeculae (Figure 12–1B).
2. The **outer cortex** contains **B cells** organized in primary and secondary lymph follicles.
3. The **paracortex** lies between follicles and contains **T cells** but lacks lymph follicles.
4. **Postcapillary venules,** also called high-endothelial venules, lie in the paracortex and have a cuboidal endothelium.
5. The walls of these venules are infiltrated with T and B cells, which pass from the blood into the lymph node.

C. Medulla

1. The **medulla,** which lies deep to the cortex and extends to the hilus, consists of medullary cords, reticular cells, reticular fibers, and sinuses.
2. **Medullary cords** are branching partitions of reticular tissue containing B and T cells, macrophages, and plasma cells.

D. Sinuses

1. **Afferent lymphatic vessels** pierce the capsule and empty into the **subcapsular sinus.**
2. **Cortical sinuses** run radially from the subcapsular sinus through the cortex and are often associated with **trabeculae.**
3. Cortical sinuses are continuous with **medullary sinuses** that converge toward the hilus and drain into the **efferent lymphatic vessels.**
4. The **lymphatic sinuses** are lined by reticular cells that extend cytoplasmic processes that cross the lumen.

LYMPHOID TISSUE TUMORS

- *Lymphomas* are tumors of lymphoid tissues that present with hard lymph nodes.
- *Hodgkin's disease* is a lymphoma characterized by chronic enlargement of lymph nodes and the presence of malignant neoplastic cells called **Reed-Sternberg cells.**
- *Non-Hodgkin's disease* is a lymphoma characterized by neoplastic transformation of B and T cells and histiocytes.

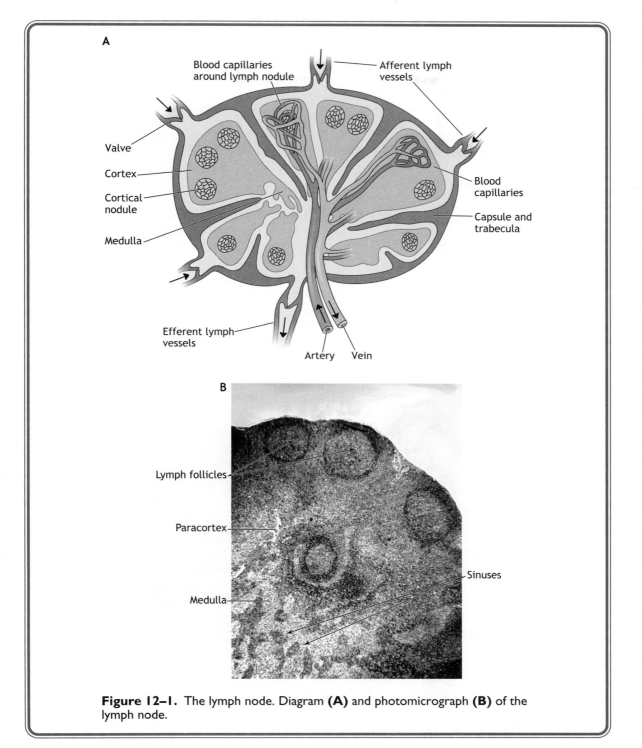

A

Blood capillaries
around lymph nodule

Afferent lymph
vessels

Valve

Cortex

Cortical
nodule

Medulla

Blood
capillaries

Capsule and
trabecula

Efferent lymph
vessels

Artery   Vein

B

Lymph follicles

Paracortex

Medulla

Sinuses

Figure 12–1. The lymph node. Diagram **(A)** and photomicrograph **(B)** of the lymph node.

V. Thymus

A. General Features

1. The **thymus** is the first lymphoid organ to develop (Figure 12–2A). It originates from both mesoderm (lymphocytes) and endoderm (epithelial cells) and thus is called a **lymphoepithelial organ.**
2. This organ produces **T lymphocytes** from precursor cells of bone marrow and is active in childhood, reaching its greatest weight at puberty.
3. The thymus consists of 2 **lobes,** each of which is composed of several **lobules.**
4. Each lobule is composed of a **medulla** and **cortex,** which lacks follicles.
5. Both the cortical and medullary regions have the same cellular components but in different proportions.

B. Function

1. The thymus functions in production of **T lymphocytes,** which circulate and populate other lymphoid organs and are involved in **cell-mediated reactions.**
2. T-cell precursors pass through the thymus before their progeny become **immunocompetent.**

C. Cortex

1. The thymic **cortex** is an active site of lymphocyte production and **maturing T lymphocytes.**
2. The blood supply to the cortex consists of capillaries.
3. Many cells die within the cortex and are phagocytized by **macrophages.**

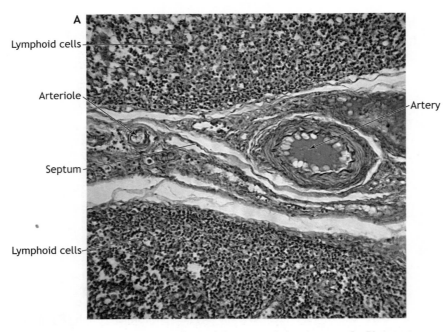

Figure 12–2. Photomicrographs and diagram of the thymus. **A:** Photomicrograph of the thymus. (continued)

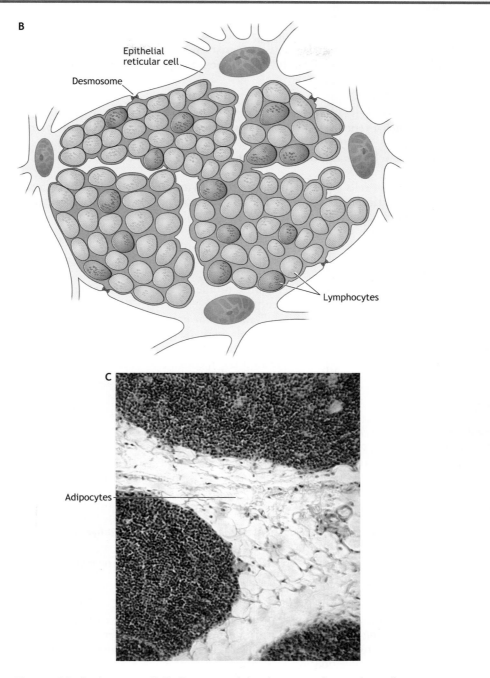

Figure 12–2. (continued) **B:** Diagram of the thymus and reticular cells.
C: Involuted thymus.

D. **Medulla**
1. The thymic **medulla** has a large number of epithelial reticular cells but few lymphocytes, which are usually mature.
2. The medulla contains **Hassall's corpuscles,** which are circular-arranged epithelial cells with no known function. The number of these structures reaches a maximum at puberty.

E. **Major Cell Types**
1. Epithelial cells (Figure 12–2B)
   a. Epithelial reticular cells form a **cytoreticulum** tightly packed with lymphocytes.
   b. These cells contain **tonofilaments** and **desmosomes** and lie on a basal lamina.
   c. Epithelial cells produce **thymosin** and **thymopoietin,** which direct the **differentiation** of **T lymphocytes.** These factors are stored in dense granules.
2. The **blood–thymic barrier** is created by reticular cells that surround blood vessels within only the **cortex.**
3. Lymphocytes
   a. Large numbers of developing **T lymphocytes** are found in the thymus before its involution.
   b. Lymphocytes in the outer cortex proliferate, become immunocompetent, and migrate into the outer cortex.
   c. T lymphocytes exposed to self-antigens or not exposed to **major histocompatibility complex (MHC)** molecules die via **programmed cell death** (apoptosis).
   d. The mature T cells migrate into the medulla and pass into the lymphatic system through **postcapillary venules.**

THYMIC INVOLUTION

- **Involution** of the thymus corresponds to a normal aging process (Figure 12–2C). During this process, cortical lymphocytes and epithelial cells are lost and replaced by **adipose tissue.**
- The thymus can undergo **stress involution,** which is related to elevated steroid hormones, such as **cortisone.** These hormones have a lytic effect on the cellular components of the thymus.

DIGEORGE SYNDROME

- **DiGeorge syndrome** is marked by underdeveloped or absent **thymus** and **parathyroid glands.** The affected individual suffers from a deficiency of T cells but not B cells.
- This disorder is a consequence of a failure of development of the third and fourth pharyngeal pouches.

VI. **Spleen**

A. **General Features**
1. The **spleen** is an immune organ that functions as a **complex filter** that modifies circulating blood and clears particulate matter and aged erythrocytes.
2. The spleen stores erythrocytes and platelets, degrades hemoglobin, and is involved in iron metabolism.

B. **Histology**
1. The spleen **does not have a cortex or medulla** but is surrounded by a thick **capsule** with many branching trabeculae (Figure 12–3A).

2. The **splenic pulp** is found within the capsule and trabeculae.
3. **Red pulp** has erythrocytes within vascular sinuses, termed **sinusoids,** and thin plates of cells, or **splenic cords.**
4. **White pulp** consists of zones of tightly packed **lymphocytes.**

C. **Splenic Arteries and Blood Flow**
1. **Splenic arteries** branch within trabeculae to form **trabecular arteries** (Figure 12–3B).
2. Trabecular arteries exit the trabeculae to enter the splenic pulp as **central arteries.**
3. Central arteries are surrounded by **periarterial lymphatic sheaths** (PALS).
4. **Collateral capillaries** supply the lymphoid tissue, terminating at the marginal zone.
5. **Penicillar arterioles** branch from the central artery.
6. **Sheathed capillaries** terminate by emptying into the splenic cords.
7. Blood flows through cords and enters the **splenic sinuses,** which form the tributaries of the veins.

D. **White Pulp**
1. The recirculating T lymphocytes and macrophages of **white pulp** are found within a reticular meshwork (Figure 12–3C).

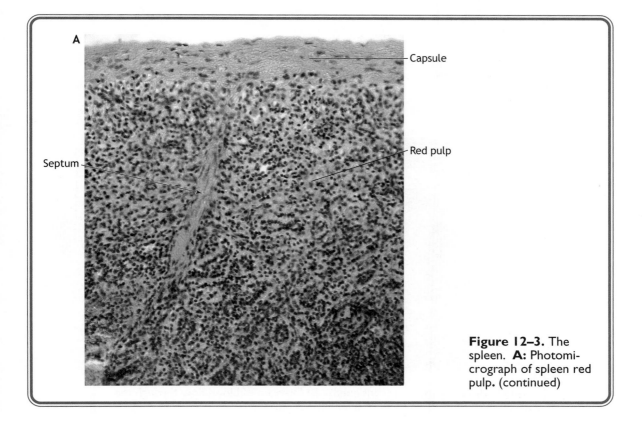

Figure 12–3. The spleen. **A:** Photomicrograph of spleen red pulp. (continued)

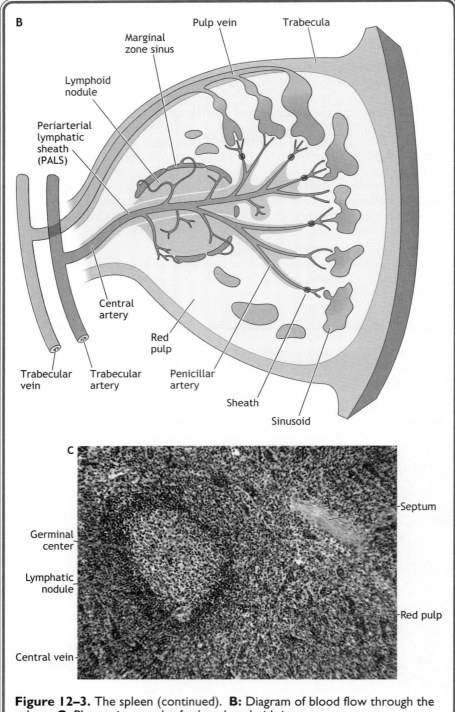

Figure 12–3. The spleen (continued). **B:** Diagram of blood flow through the spleen. **C:** Photomicrograph of spleen lymphoid tissue.

2. The lymphatic nodules within the **PALS** may contain germinal centers that are concentrated with B lymphocytes.

3. The **marginal zone** between the white and red pulp contains **antigen-presenting cells,** primarily **macrophages,** which search for antigens in the passing blood within the sinuses.

E. Red Pulp

1. The **red pulp** is a reticular meshwork supplied by arteries and drained by venous sinuses.

2. This region is a branching system of cords lying between sinuses called **splenic cords.**

3. Arterial vessels open into the reticular meshwork of the cords, producing an open circulation.

4. **Sheathed capillaries** are surrounded by **macrophages,** which phagocytize damaged cells.

F. Splenic Sinuses

1. **Splenic sinuses** have a unique endothelium and basement membrane.

2. The endothelial cells are separated by slitlike spaces, and the basement membrane is perforated by **fenestrae.**

3. Blood cells cross this wall and enter the sinus.

4. **Splenic sinuses** are tributaries of **pulp veins** that flow into trabecular veins.

ENLARGEMENT OF THE SPLEEN

• *Enlargement of the spleen, or **acute splenic tumor,** results from an infection of the blood elements, leading to congestion within the red pulp. Neutrophils and plasma cells accumulate within the white and red pulp.*

CONGESTIVE SPLENOMEGALY

• ***Congestive splenomegaly*** *can be caused by cirrhosis of the liver and venous congestion related to portal or splenic drainage or veins within the substance of the liver.*

CLINICAL PROBLEMS

1. Hassall's corpuscles are found in which of the following regions of the thymus?

A. Medulla

B. Thymic lobules

C. Septal connective tissue

D. Cortical-medullar junction

E. Cortex

2. Periarterial lymphatic sheaths in the spleen surround which of the following vessels?

A. Splenic sinuses

    B. Penicillar arterioles

    C. Central arterioles

    D. Splenic arteries

    E. Collateral capillaries

3. Lymph is carried to a lymph node via afferent lymphatic vessels. These vessels drain directly into which of the following?

    A. Cortical sinuses

    B. Medullary sinuses

    C. Trabecular sinuses

    D. Subcapsular sinuses

    E. Efferent lymphatic vessels

A young African-American girl is diagnosed with homozygous sickle cell anemia. She is anemic and displays a large population of sickled-appearing erythrocytes within her peripheral blood.

4. In which of the following sites will the abnormal erythrocytes be removed from the circulation?

    A. Thymic cortex

    B. Capsular region of spleen

    C. Nodular region of lymph node

    D. Thymic medulla

    E. Billroth's cord of the spleen

Your patient presents with seizures and suffers frequent episodes of immune-associated disorders. You diagnose him as having DiGeorge syndrome.

5. Which of the following organs would be underdeveloped in this individual?

    A. Liver

    B. Thymus

    C. Spleen

    D. Thyroid gland

    E. Pancreas

6. Which of the following histologic features best distinguishes the spleen from the lymph node?

    A. Lack of nodular lymphoid tissue

    B. Sinuses

    C. No cortex or medulla

    D. B and T cells

    E. Diffuse lymphoid tissue

7. Which of the following vessels within the spleen drain directly into splenic cords?

    A. Sheathed capillaries

    B. Splenic sinuses

    C. Central arteries

    D. Penicillar arterioles

    E. Trabecular arteries

While examining a lymphoid tissue, a pathologist notes a distinctive cortex and medulla. In addition, vessels are observed entering at the capsule and leaving at a hilus.

8. Which of the following tissues is being examined?

    A. Spleen

    B. Lymph node

    C. Thymus

    D. Peyer's patch

    E. Encapsulated lymphoid nodule

9. Within a spleen, which of the following cells is an example of an antigen-presenting cell?

    A. B lymphocyte

    B. Neutrophil

    C. Plasma cell

    D. T lymphocyte

    E. Macrophage

ANSWERS

1. The answer is E. Hassall's corpuscles, also called thymic corpuscles, are found within the cortex of the thymus.

2. The answer is C. Central arterioles (and sometimes arteries) are surround by lymphoid cells, called the periarterial sheath.

3. The answer is D. Lymph is transported to lymph nodes via afferent lymphatic vessels. These vessels pierce the lymph node capsule and drain into subcapsular sinuses.

4. The answer is E. Damaged or sickled erythrocytes are removed from circulation within Billroth's cords. These cells are also removed within the bone marrow.

5. The answer is B. DiGeorge syndrome is characterized by underdevelopment or absence of the thymus and parathyroid glands.

6. The answer is C. The spleen does not have a cortex or medulla but consists of red and white pulp.

7. The answer is A. Sheathed capillaries drain directly into splenic cords.

8. The answer is B. Lymph nodes have both a cortex and medulla. Afferent lymphatic vessels empty into subcapsular sinus, whereas efferent vessels exit the node at the hilus.

9. The answer is E. Macrophages are the principal antigen-presenting cell within the spleen. Other cells that present antigens to T cells include Langerhans' cells of the skin and B cells and reticular cells within the thymus.

CHAPTER 13
INTEGUMENT

I. Functions

A. The **integument,** which means covering, protects the body against bacteria, sunlight, mechanical forces, dehydration, and cold.

B. Other functions of the integument include metabolism of fat, formation of **vitamin D** by the action of ultraviolet light, sensory appreciation of the environment by specific **receptors,** and **friction** as in grasping.

C. The integument also regulates **temperature** by varying peripheral blood flow, sweating, hair elevation, and **insulation** provided by adipose tissue.

II. Epidermis

A. **General Features**
   1. The **epidermis** consists of a stratified squamous keratinized epithelium with keratinocytes, melanocytes, Langerhans' cells, Merkel's cells, and mitotic cells (Figure 13–1A and B).
   2. **Thick skin** is found on the palm and soles (Figure 13–1C), whereas **thin (hairy) skin** is located elsewhere in the body (Figure 13–1D).

B. **Stratum Basale**
   1. The **stratum basale** is the deepest layer of the epidermis and consists of a single layer of cuboidal or columnar cells that lie on a basal lamina.
   2. Cells of this layer are bound together by **desmosomes,** whereas **hemidesmosomes** attach these cells to the basal lamina.
   3. Cells undergoing **mitosis** are typical of this layer, which provides constant renewal of epidermal cells every 15–30 days.
   4. Cells in this layer contain **cytokeratins,** which are a type of intermediate filament protein. The content of this protein increases in cells of the upper layers of the epidermis.

C. **Stratum Spinosum**
   1. Cells of the **stratum spinosum** are cuboidal, polygonal, or flattened.
   2. Cells in this layer are interconnected by **desmosomes,** giving a spiny appearance after tissue preparation.
   3. Tonofibrils, bundles of **tonofilaments,** insert into the cytoplasmic densities of the desmosomes and function to maintain cell attachment and provide resistance to abrasion.
   4. The **malpighian layer,** which includes the strata basale and spinosum, contains mitotic cells.

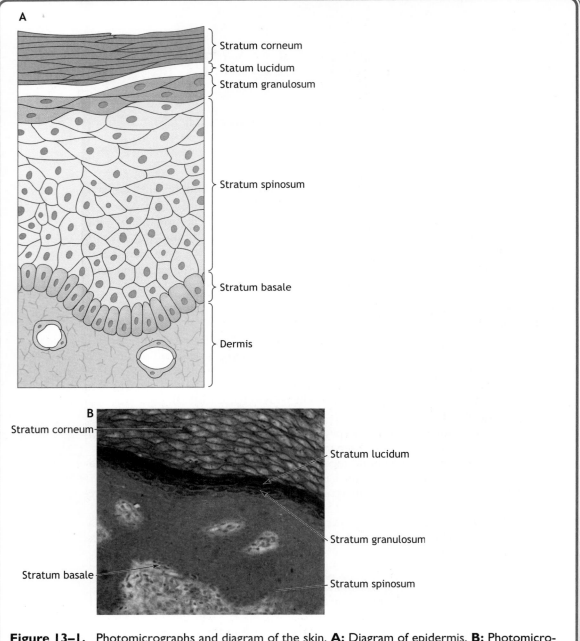

Figure 13–1. Photomicrographs and diagram of the skin. **A:** Diagram of epidermis. **B:** Photomicrograph of epidermis. (continued)

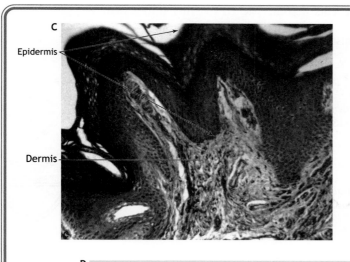

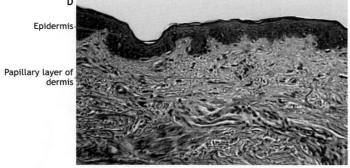

Figure 13–1. (continued)
C: Photomicrograph of thick skin.
D: Photomicrograph of thin skin.

D. Stratum Granulosum

1. The stratum granulosum consists of 3–5 layers of flattened polygonal cells.
2. These cells are filled with **keratohyalin granules,** which are basophilic and contain phosphorylated histidine-rich, cystine-containing proteins.
3. **Lamellar granules** in these cells release a **glycolipid** by exocytosis. This material provides a sealing property to skin and a barrier to penetration by foreign materials.

E. Stratum Lucidum

1. The **stratum lucidum** is a translucent, thin layer that is most apparent in thick skin.
2. This layer consists of flattened **eosinophilic cells** joined by desmosomes in which organelles and nuclei are no longer present.
3. The cytoplasm consists of densely packed filaments.

F. Stratum Corneum

1. The **stratum corneum** consists of 15–20 layers of flattened keratinized cells, which do not have nuclei. In thin skin, this layer is thinner.
2. **Keratin,** a filamentous protein containing 6 different polypeptides (40–70 kDa), fills the cytoplasm of these cells.

3. In cells of this layer, tonofilaments become packed together into a matrix with keratohyalin granules.
4. Cells of this layer are continuously shed at the surface, and renewal occurs approximately every 4 weeks.

PSORIASIS

- *Psoriasis is a skin disease in which mitotic activity of the strata basale and spinosum is increased above normal but cell cycle time is decreased. Thus, renewal of cells of the epidermis may occur every 7 days instead of 4 weeks.*

SKIN TUMORS

- *Most of tumors of the skin are derived from basal cells, which produce **basal cell carcinomas,** and cells of the stratum spinosum, which produce **squamous cell carcinomas.***

III. Cells of Integument

    A. Melanocytes

        1. **Melanocytes** are **neural crest-derived cells** found within the stratum basale and hair follicles.

        2. These cells are bound to the basal lamina by **hemidesmosomes.**

        3. Extensions of melanocytes branch into the strata basale and spinosum.

        4. The tips of these extensions extend into invaginations of cells in these strata.

        5. These cells are pale staining but have abundant mitochondria, Golgi complexes, and rough endoplasmic reticulum (RER).

MELANOCYTE TUMORS

- *The **malignant melanoma** is an invasive tumor of melanocytes and accounts for 1–3% of all tumors. This tumor consists of rapidly dividing, malignantly transformed melanocytes of the **stratum basale.***
- *These malignant cells penetrate the basal lamina, pass into the dermis, and enter the blood vessels to gain access to the rest of the body.*

    B. Langerhans' Cells

        1. **Langerhans' cells** are found primarily in the **stratum spinosum** and account for 3–8% of epidermal cells.

        2. These cells are actually bone marrow-derived **macrophages** that are capable of binding, processing, and presenting antigens to T lymphocytes; thus, these cells play a role in immunologic skin reactions.

    C. Merkel's Cells

        1. **Merkel's cells** are found primarily in thick skin of the palm and sole in the **stratum basale.**

        2. Free nerve endings are found at the base of these cells, indicating that these cells may serve as a sensory mechanoreceptor.

IV. Melanin Synthesis

    A. Synthesis of melanin by **melanocytes** is dependent on **tyrosinase** activity.

    B. **Tyrosine** is converted to 3,4-dihydroxyphenylalanine (dopa) and then to dopaquinone. Ultimately, through a series of transformations, melanin is produced.

C. Within the melanocyte, **tyrosinase** is made on ribosomes, transported to the lumen of RER, and accumulates within vesicles of the Golgi complex.

D. The **melanin granules** migrate within the cytoplasmic extension of the melanocyte and then are transported into keratinocytes of the strata basale or spinosum, a process described as **cytocrine secretion.**

E. The melanin granules ultimately become oriented in the **supranuclear region,** thus protecting nuclei from the effects of solar radiation.

F. Melanin disappears from cells of the upper layers of skin because the granules fuse with **lysosomes** and are broken down.

MELANIN DISORDERS

- *Albinism* is a hereditary disease in which melanocytes have a decreased capacity to synthesize melanin. This is caused by an absence of tyrosinase activity or an inability of cells to take up tyrosine.
- *Reduced cortisol* release from the adrenal cortex causes increased production of adrenocorticotropic hormone (ACTH). This leads to increased pigmentation of the skin. In *Addison's disease,* dysfunctional adrenal cortex results in increased skin pigmentation.
- *Tanning* (ie, exposure to ultraviolet rays of the sun) causes a darkening of the preexistent melanin, which is then released into the keratinocyte. The rate of melanin synthesis is also accelerated, thereby increasing the level of melanin.

V. Dermis

A. **General Features**

1. The **dermis,** which consists of a superficial **papillary layer** and a deeper **reticular layer** of connective tissue, supports the epidermis and binds it to the hypodermis (Figure 13–2).

2. The dermis varies in thickness but may be as thick as 5 mm on the back.

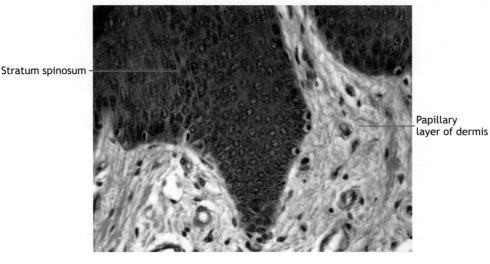

Figure 13–2. Photomicrograph of dermis.

    3. Dermal papillae interdigitate with projections of the epidermis that are called **epidermal ridges** or rete pegs.

    4. A basal lamina is found between the stratum basale and dermis.

  B. Blood and Nervous Supply

    1. The dermis is rich in blood and lymph vessels.

    2. Dermal vessels account for **4.5% of the blood supply** at any one time.

    3. The capillary network in the papillary layer surrounds the epidermal ridges and regulates body temperature and nourishment of the epidermis, which lacks blood vessels.

    4. The dermis has a supply of nerves from sympathetic ganglia of the paravertebral chain.

VI. Hypodermis

  A. The hypodermis consists of loose connective tissue and contains **adipocytes.**

  B. This layer is also called **superficial fascia** or, if thick, the **panniculus adiposus.**

DISORDERS OF THE HYPODERMIS

- An inflammatory reaction of the subcutaneous fat is called **panniculitis.** The most common form is **erythema nodosum,** caused by infections and drugs. Less common, **erythema induratum** is associated with subdural vessels

VII. Hair

  A. General Features

    1. A **hair** is an elongated keratinized structure that is derived from invaginations of the epidermal epithelium (Figure 13–3).

    2. Hairs are found everywhere on the body except areas such as soles of the feet and palms of the hands.

    3. Hair growth occurs at the **hair bulb** and is a discontinuous process that is influenced by sex hormones, androgens, and adrenal and thyroid hormones.

  B. Hair Follicle

    1. The **hair follicle** arises from the epidermal invagination and during its growth period has a hair bulb.

    2. The central region of the follicle is the **medulla,** which consists of cells that are moderately keratinized.

    3. The cells surrounding the medulla are the hair **cortex,** which are fusiform and heavily keratinized.

    4. The hair **cuticle** is peripheral to the cortex and consists of a layer of cells that are cuboidal up to the middle of the follicle and then become tall and columnar. Higher up, the cells become vertical, and they then form a layer of flattened, heavily keratinized cells covering the cortex.

    5. The **internal root sheath** (IRS) is lost above the level of the sebaceous gland.

    6. The **external root sheath** (ERS) is continuous with epidermal cells.

    7. The **glassy membrane** separates the hair follicle from the dermis. This membrane is actually a basal lamina.

    8. The **connective tissue sheath** surrounds the glassy membrane. The **arrector pili muscle,** consisting of smooth muscle, connects the connective tissue sheath to the papillary layer of the dermis (Figure 13–4).

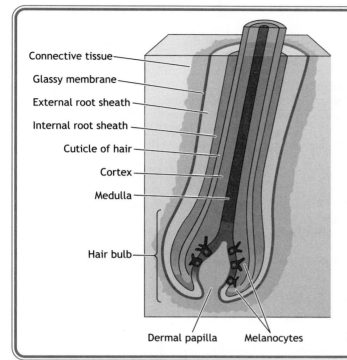

Connective tissue

Glassy membrane

External root sheath

Internal root sheath

Cuticle of hair

Cortex

Medulla

Hair bulb

Dermal papilla   Melanocytes

Figure 13–3. Diagram of a hair follicle.

9. Contraction of the arrector pili muscle leads to a depression of the skin, which produces **goose flesh.**

C. **Hair Color**

1. The **color of hair** is caused by **melanocytes** found in the cortex of the hair follicle.
2. Melanocytes produce and transfer melanin to epithelial cells via a mechanism similar to that in the epidermis.
3. Injury to the dermal papillae leads to loss of hair.

VIII. **Glands of the Skin**

A. **Sebaceous Glands**

1. **Sebaceous glands** are found in the dermis over most of the body (Figure 13–4).
2. These are acinar glands, with several acini that open into a small duct, which opens into the upper portion of the **hair follicle.**
3. As acinar cells differentiate, they fill with fat droplets and ultimately burst, and the sebum gradually moves to the surface of the skin.
4. This gland is a type of **holocrine gland** because its secretion is released with remnants of the cell.
5. The function of sebum may have some antibacterial or antifungal properties.

SEBACEOUS GLAND DISORDER

• *Sebum from the sebaceous gland flows continuously, but if this flow or secretory process is disturbed **acne** develops. Acne results from a chronic inflammation of the obstructed sebaceous gland.*

B. **Sweat Gland**
  1. Eccrine sweat glands
     a. **Eccrine glands** are simple, coiled tubular glands, whose ducts **open at the surface of the skin** (Figure 13–4).
     b. The **secretory segment** is found in the dermis, and surrounding **myoepithelial cells** aid in the secretion process.
     c. The **secretory ducts** are lined by stratified cuboidal epithelium.
     d. The fluid secreted by these glands is not viscous but consists of water, sodium chloride, and urea.
     e. On the skin surface, sweat evaporates, cooling the surface.
  2. Apocrine sweat glands
     a. **Apocrine glands** are larger than eccrine sweat glands, and their **ducts open into the hair follicle.**
     b. The **viscous secretion** is initially odorless, but with bacterial decomposition it acquires its distinctive odor.
     c. These glands are innervated by **adrenergic** nerve endings.

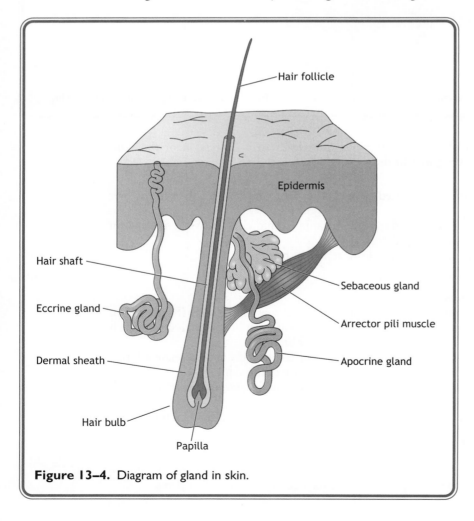

Figure 13–4. Diagram of gland in skin.

CLINICAL PROBLEMS

1. Langerhans' cells are found in which of the following layers of the epidermis?
   A. Stratum basale
   B. Stratum spinosum
   C. Stratum corneum
   D. Stratum lucidum
   E. Stratum granulosum

2. Which of the following separates the hair follicle from the connective tissue of the dermis?
   A. External root sheath
   B. Internal root sheath
   C. Glassy membrane
   D. Hair cuticle
   E. Medulla

3. Which of the following layers is not found in thin skin?
   A. Stratum basale
   B. Stratum spinosum
   C. Stratum granulosum
   D. Stratum lucidum
   E. Stratum corneum

4. Which of the following components of the epidermis provide sealant between adjacent cells?
   A. Glycolipids
   B. Keratin
   C. Keratohyalin granules
   D. Desmosomes
   E. Tight junctions

Your patient presents with patchy lesions of the skin. You observe sloughing of surface cells of the epidermis.

5. Which of the following would cause this clinical condition?
   A. Loss of tight junctions in upper skin layers
   B. Longer cell cycle of cells in the deep skin layers
   C. Increased mitosis of cells of the stratum basale
   D. Reduction of keratin synthesis
   E. Decreased mitosis of cells of the malpighian layer

6. Which of the following best characterizes sebaceous glands?

    A. Its duct drains onto the skin surface

    B. It releases it contents via holocrine secretion

    C. Its secretion is primarily water and salts

    D. Its secretory units are supplied by adrenergic stimulation

    E. Its myoepithelium aids in the secretion process

A young boy walking in the woods scrapes his leg on a shrub. He suffers a rash that spreads overnight, but with application of ointment the next day the rash disappears.

7. Which of the following cells was involved in this rash?

    A. Keratinocytes

    B. Melanocytes

    C. Mitotic cells

    D. Merkel's cells

    E. Langerhans' cells

An extremely small sample of hairy skin is found at a crime scene and brought into the police laboratory. It is determined that the sample size is not adequate for proper analysis. The investigators want to expand the number of cells for analysis by restriction fragment length polymorphism (RFLP).

8. Cells from which of the following can be expanded to provide a sufficient cell sample for this analysis?

    A. Hair bulb of hair follicle

    B. External root sheath of hair follicle

    C. Stratum lucidum

    D. Cuticle of hair shaft

    E. Stratum granulosum

A 2-year-old boy suffers an allergic contact dermatitis on the dorsum of his hand. His pediatrician applies an ointment to the patchy skin.

9. The physician tells the mother that she can expect improvement in the skin within which of the following time periods?

    A. One week

    B. Two weeks

    C. Four weeks

    D. Six weeks

    E. Eight weeks

10. Myoepithelial cells aid in the secretory process of which of the following?

    A. Sebaceous glands

    B. Eccrine glands

C. Keratinocytes

D. Melanocytes

E. Apocrine glands

ANSWERS

1. The answer is B. Langerhans' cells, also called dendritic cells, are found within the stratum spinosum of the epidermis.

2. The answer is C. Hair follicles are separated from the connective tissue of the dermis by the glassy membrane, which is actually a basal lamina.

3. The answer is D. Thin skin usually does not have the stratum lucidum. Some cells of this layer, however, may be found in its place.

4. The answer is A. Glycolipid is released from granules of cells within the stratum granulosum. This lipid-rich substance acts as a sealant to provide a watertight seal between cells. Keratin provides rigidity to keratinocytes themselves, whereas desmosomes interconnect with cells but provide no barrier between cells.

5. The answer is C. The skin condition described is psoriasis, which is a disease marked by increased mitotic activity of cells in the stratum basale.

6. The answer is B. Sebaceous glands release the total contents of the cell and thus secrete by holocrine secretion. The duct of this gland drains into the hair follicle. Its secretion is sebum and thus has a mucoid consistency.

7. The answer is E. Langerhans' cells are found within the stratum spinosum and are macrophages that are antigen-presenting cells.

8. The answer is A. The hair bulb of the hair follicle contains mitotically active cells. This region is equivalent to the stratum basale or germinativum of skin. The strata lucidum and granulosum do not have mitotic cells. Similarly, mitotic cells are not found within the external root sheath or cuticle.

9. The answer is C. Renewal of skin occurs over a period of 4 weeks. A keratinocyte originating in the stratum basale after mitosis will migrate to the surface epithelium within a 4-week period.

10. The answer is B. Myoepithelial cells surround the secretory element of eccrine sweat glands, which aid in discharge of the contents of secretory granules. Secretion by sebaceous glands is facilitated by contraction of the arrector pili muscle through a duct. Melanocytes transfer melanin granules into keratinocytes by the pinching off of the plasma membrane of its process containing the granule. This process is called cytocrine secretion. Apocrine sweat glands secrete via a duct on the skin surface.

CHAPTER 14
GASTROINTESTINAL SYSTEM

N

I. General Features

A. The gastrointestinal (GI) system consists of 3 major parts.
   1. The **oral cavity** includes salivary glands, tongue, and teeth.
   2. The **tubular tract** consists of the esophagus, stomach, small intestine, large intestine, rectum, anal canal, and anus.
   3. The **accessory GI glands** include the liver, gallbladder, and exocrine pancreas.

B. The **functions** of the GI system include **digestion** of food material to a small particulate size, **hydrolysis** of complex food into simple residues, and **absorption** of these residues.

C. The wall of the digestive tube consists of a **mucosa** composed of an epithelium, lamina propria, and a muscular component, a smooth muscle layer, and a serosa layer (Figure 14–1).

II. Salivary Glands

A. **General Features**
   1. The bulk of **saliva** comes from the parotid, submandibular, and sublingual glands.
   2. Saliva, consisting of mixed secretions from all salivary glands, is secreted on mechanical, chemical, olfactory, or psychic stimuli.
   3. **Salivary secretions** of 1–1.5 L every day contain water, mucus, proteins, salts, cells, and enzymes.
   4. Saliva primarily acts to moisten the oral cavity, clean the mouth, lubricate food, assist in the taste sensation, and digest carbohydrates.

B. **Parotid Gland**
   1. **Parotid glands,** the largest of the salivary glands, are classified as branched acinar glands and produce 25% of total saliva.
   2. The fibrous **capsule** that surrounds the parotid gland contains plasma cells and lymphocytes and is divided into **lobes** by connective tissue **septa.**
   3. These glands are composed almost exclusively of **serous secretory cells** with secretory granules rich in proteins and enriched in **amylase.**

PAROTID GLAND DISEASE

· *Swelling of the parotid gland resulting from a viral infection is called **mumps,** a common childhood diease. The resulting inflammation can affect the parotid glands and, less frequently, the CNS and pancreas.*

CLINICAL CORRELATION

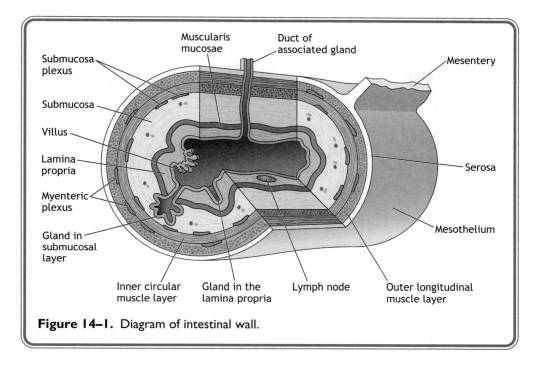

Figure 14–1. Diagram of intestinal wall.

C. **Submandibular (or Submaxillary) Gland**
1. The **submandibular gland** is a branched tubuloalveolar gland consisting of both **serous** and **mucous** secretory units that produce about 70% of saliva.
2. The majority of the secretory units (75%) are **serous** acini.
3. The mucous units can be capped with **serous demilunes.**
4. The serous cells of the serous demilunes secrete **lysozyme.**

D. **Sublingual Glands**
1. **Sublingual glands** are not well encapsulated but are a collection of glands lying beneath the mucous membrane at the floor of the mouth.
2. These glands are classified as a branched tubuloalveolar gland formed by both **serous** and **mucous** secretory units.
3. The majority of the secretory units are **mucous** (75%).

E. **Duct System**
1. Secretions of salivary glands first enter an **intercalated duct,** which is lined by a cuboidal epithelium. They then pass into **striated ducts** lined by columnar epithelium.
2. Cells of **striated ducts** have an acidic cytoplasm, basal membrane invaginations, and abundant **mitochondria,** which are characteristics of ion-transporting cells.

III. **Tongue**

A. **General Features**
1. The **tongue** is covered by a mucous membrane that is rough on the dorsal side and smooth on the ventral side.

2. The rough surface is due to small protuberances called **papillae.**

3. The tongue consists of interlacing bundles of skeletal muscle at various angles.

4. The posterior one third of the dorsal surface of the tongue is separated from the anterior two thirds by the V-shaped **sulcus terminalis.**

B. Types of Papillae of Tongue

   1. Filiform papillae

      a. Filiform papillae have an elongated conical shape and consist of a connective tissue core covered by an epithelial layer of stratified squamous cells.

      b. These papillae are found in rows parallel to the median sulcus but do not contain taste buds.

   2. Fungiform papillae

      a. Fungiform papillae are mushroom shaped and contain a few taste buds on their upper surface.

      b. These papillae are scattered among filiform papillae and are most numerous near the tip of the tongue.

      c. Fungiform papillae are larger and fewer in number than filiform papillae.

   3. Circumvallate papillae (Figure 14–2A)

      a. Circumvallate papillae are located along the sulcus terminalis and number from 7–12.

      b. These papillae are surrounded by a deep, circular, moatlike furrow. **Taste buds** are located in its lateral surface.

      c. Serous Ebner's glands drain a watery fluid containing lipases into the furrow of these papillae.

C. Taste Buds

   1. Taste buds are pale, barrel-shaped **intraepithelial bodies** that extend from the basal lamina to the surface (Figure 14–2B).

   2. Each taste bud consists of 40–70 cells, including a **chemoreceptive neuroepithelial** cell, support cells, and basal cells.

   3. The taste buds have a small opening, called the **taste pore,** and receive innervation at their base.

IV. Esophagus

A. General Features

   1. The **esophagus** is a straight tube about 8 in long that extends from the pharynx to the stomach (Figure 14–3A).

   2. Food passes very rapidly through the esophagus.

B. Components of the Wall of the Esophagus

   1. The esophagus is lined by a nonkeratinized, stratified squamous epithelium, which undergoes continual renewal.

   2. The **muscularis mucosa** is thickest in the esophagus, consisting of 20 layers.

   3. The **submucosa** contains coarse collagenous and elastic fibers and is thrown into folds.

   4. The upper third of the **muscularis externa** contains skeletal muscle associated with the pharynx; the middle third contains both skeletal and smooth muscle; and the lower third consists of only smooth muscle.

C. Only that portion of the esophagus that is found in the peritoneal cavity is covered by a serosa; the rest is covered by an adventitia. The adventitia blends into the surrounding tissue.

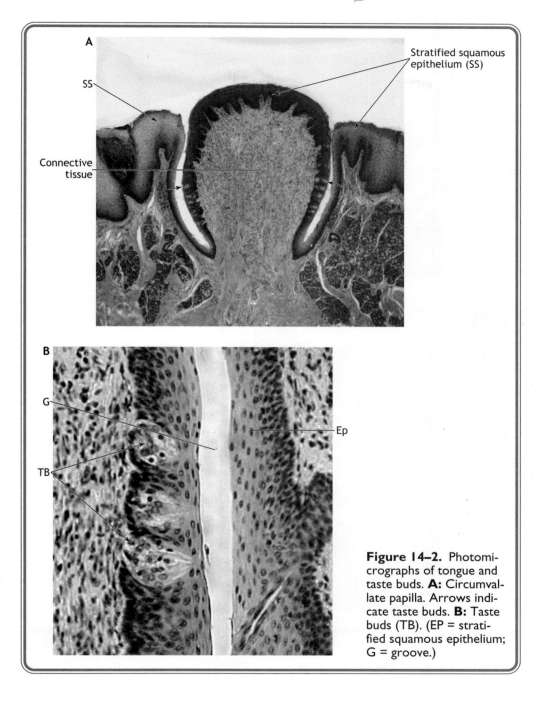

Figure 14–2. Photomicrographs of tongue and taste buds. **A:** Circumvallate papilla. Arrows indicate taste buds. **B:** Taste buds (TB). (EP = stratified squamous epithelium; G = groove.)

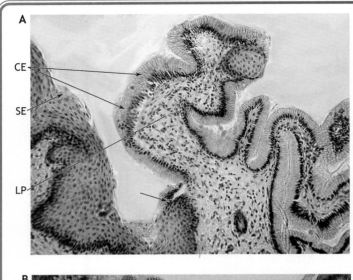

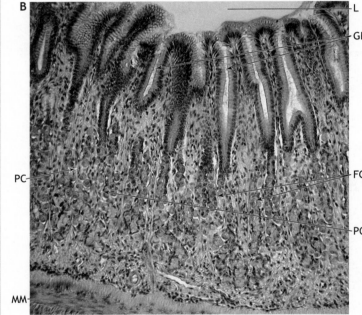

Figure 14–3. Photomicrographs of esophagus and stomach. **A:** Esophagus. Arrow indicates apparent transition of the stratified squamous epithelium of the esophagus (SE) to the simple columnar epithelium (CE) of the stomach. (LP = lamina propria.) **B:** Fundic stomach. (L = lumen; GP = gastric pits; FG = fundic gland; PC = parietal cell; MM = muscularis mucosae.)

D. Glands
1. A few mucous glands are scattered in the submucosa and are called **esophageal glands.**
2. Mucous glands near the stomach in the lamina propria are called **esophageal cardiac glands.**

BARRETT'S ESOPHAGUS

- *Extensive gastroesophageal reflux leads to **Barrett's esophagus.** The stratified squamous epithelium covering the distal esophagus is transformed to a simple columnar epithelium. These patients are at high risk for **adenocarcinomas.***

SQUAMOUS CELL CARCINOMA

- *A high percentage of the malignant cancers (90%) of the esophagus are **squamous cell carcinomas.** The most common benign tumor of the esophagus is the **leiomyoma,** which is derived from **smooth muscle** of the esophageal wall.*

V. Stomach

    A. General Features

        1. The **stomach** is a dilated segment of the digestive tract that acts as a reservoir for ingested food, which is termed **chyme** (Figure 14–3B).

        2. Digestion is initiated in the stomach by the action of enzymes, such as **pepsin** and **lipase.**

        3. Rugae are folds that greatly increase the internal surface area of the stomach.

    B. Components of the Wall of the Stomach

        1. Epithelium

            a. The surface epithelial layer consists of simple columnar epithelium, which is replaced every **4–7 days.** Cells die via apoptosis and are sloughed off at the luminal surface.

            b. Mucin provides a protective coating over this epithelium.

            c. Gastric glands form from surface invaginations. These glands are simple branched tubular glands located in the lamina propria that open via gastric pits.

        2. The **lamina propria,** most obvious in the cardiac and pyloric regions consists of collagen and reticular fibers and lymphocytes, mast cells, and plasma cells.

        3. The **muscular mucosa** consists of slips of smooth muscle extending into the lamina propria between gastric glands.

        4. The **submucosa** consists of dense connective tissue and collagen, reticular, and elastic fibers. The submucosa contains blood vessels, lymph vessels, **submucosal plexus of Meissner,** and cells, including macrophages, mast cells, and lymphoid cells.

        5. The **muscular externa** has **3 layers of smooth muscle** arranged as outer longitudinal, middle circular, and inner oblique layers. The **myenteric plexus of Auerbach** is found between the circular and longitudinal layers, and its major function is to stimulate the mixing action of the stomach.

        6. The **serosa** consists of a layer of loose connective tissue, blood vessels, and nerves covered by a mesothelium.

    C. Cardiac Region and Its Glands

        1. The **cardiac region** of the stomach is a narrow circular band, only 1.5–3 cm wide, which is found between the esophagus and stomach.

        2. Cardiac glands

            a. Cardiac glands are found at the esophageal orifice and are simple or branched tubular glands that extend into the deeper half of the mucosa.

            b. These glands resemble glands of the terminal region of the esophagus.

      c. These short glands have short pits but a long glandular part.

      d. Columnar cells, a major cell of this gland, secrete mucus and lysozyme.

D. Fundus and Body of Stomach

    1. The **fundus** is a dilated region that lies to the left of and superior to the cardiac orifice. The largest portion of the stomach is the body.

    2. **Fundic glands** are branched, tubular glands found in the lamina propria that make up more than 99% of stomach glands (Figure 14–3B).

E. Cells of Fundic Gland

    1. Surface **epithelial cells** produce a basic mucus, which has a protective role.

    2. **Undifferentiated cells** undergo mitosis to replace cells of the gland. These cells are columnar and are found in the neck region. Some of these cells migrate up toward the lumen and replace surface mucous cells, which have a turnover rate of **4–7 days,** although other glandular cells are replaced more slowly.

    3. **Chief cells** are basophilic cells found in the lower region of the tubular gland. These cells produce enzymes such as **lipase** and **pepsinogen,** which is converted to pepsin in the acidic environment of the stomach. These cells have apical granules.

    4. **Parietal cells** are found at the upper region of the gastric gland but are sparse at the base. These cells are pyramidal or round, with a central nucleus and eosinophilic cytoplasm. Parietal cells have numerous **mitochondria** and deep invaginations of apical membrane forming canaliculi. When inactive, parietal cells have tubulovesicles at their apical region. However, when stimulated to produce **HCl,** these vesicles fuse with the cell membrane to form microvilli, which increases the surface area. These cells are the source of HCl, KCl, and **intrinsic factor. Gastrin** and **histamine** secreted by the gastric mucosa act to stimulate HCl production. Hydrogen ions are produced in parietal cells by the dissociation of H_2CO_3 by carbonic anhydrase.

    5. **Mucous neck cells,** found between parietal cells in the neck of gastric glands, produce mucus. This **acid mucopolysaccharide** may protect the gland from the effects of the acidic environment. These cells are irregular in shape, with nuclei at their base.

    6. **Enteroendocrine cells** are located at the base of glands and have basal granules. These cells produce **gastrin, histamine, cholecystokinin,** or **serotonin.** In **closed-type cells,** the cell apex is covered by epithelial cells, and the secretory products of these cells pass into the lamina propria, where they are then picked up by the capillary network within this region. In **open-type cells,** the apex of the cell has microvilli and directly contacts the lumen. It is thought that the luminal contents act on these microvilli, stimulating the secretory activity of these cells.

F. Pylorus and Its Glands

    1. The body of the stomach continues into a region called the **pyloric antrum,** which narrows to the pyloric canal, which further narrows to the pylorus.

    2. The **pylorus** has deep gastric pits into which the branched tubular pyloric glands open. These glands are relatively short.

    3. **Pyloric glands** consist primarily of mucus-producing cells, although some lysozyme-secreting cells are found.

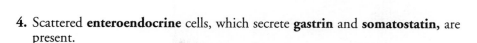

4. Scattered **enteroendocrine** cells, which secrete **gastrin** and **somatostatin,** are present.

GASTRITIS

* *Inflammation of the mucosa of the stomach is termed **gastritis,** which may be acute or chronic.*
* ***Acute gastritis** is characterized by **neutrophils** invading the epithelium, with loss of most of the superficial surface epithelium and possibly hemorrhages.*
* ***Chronic gastritis** is characterized by epithelial metaplasia, extensive mucosal atrophy with infiltration of **lymphocytes,** and plasma cells within the lamina propria.*

PEPTIC ULCERS

* *Chronic lesions, or **peptic ulcers,** can occur in the stomach mucosa as well as other regions of the gastrointestinal tract as a result of exposure to the acidic environment.*
* *The first part of the duodenum and the stomach are the first and second most common GI tract sites, respectively, of peptic ulcers.*

CARCINOMA

* *The **carcinoma** is the most common malignant tumor of the stomach. These tumors may project into the lumen or extend as a cavity into the mucosa and underlying lamina propria and muscularis.*
* *A **polyp** is usually a nonneoplastic growth that extends above the mucosa, including the lamina propria.*

VI. Small Intestine

A. General Features
1. The **small intestine** is ~16 feet long. This length permits prolonged contact between ingested food materials and the digestive enzymes as well as the products of digestion and the absorptive cells of the small intestine (Figure 14–4A).
2. Breakdown of food is completed in the small intestine.
3. Absorption of products into blood and lymph vessels, called **lacteals,** occurs within the small intestine.

B. Epithelial Specialization
1. The small intestine has a large number of **absorptive epithelial cells.**
2. The **mucous membrane** is thrown into circular folds called **plicae circulares** (Kerckring's valves), which increase the surface area **3-fold.** The plicae circulares consists of the mucosa and submucosa and is most prominent in the **jejunum,** although it may be found in the duodenum and ileum.
3. **Villi** are finger-like projections consisting of mucosa and lamina propria that increase the surface area **30-fold.**
4. Villi are covered with absorptive cells and **goblet cells.**
5. Between villi are intestinal glands, also called **glands of Lieberkühn.**
6. As many as 3000 **microvilli** project from the luminal surface of absorptive cells. These structures greatly increase the contact between the intestine and food material and increase the surface area by **600-fold.**

C. Lipid Absorption
1. Most lipids are absorbed in the duodenum and upper jejunum.

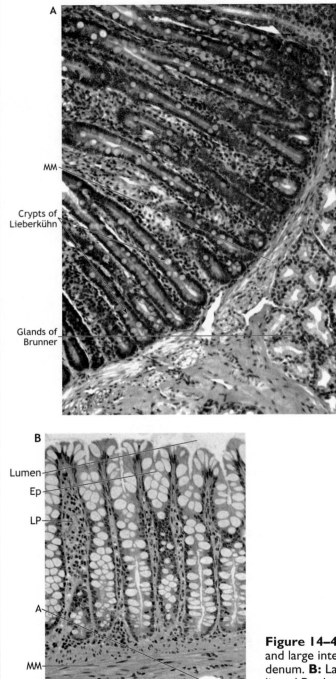

A

MM

Crypts of
Lieberkühn

Glands of
Brunner

B

Lumen

Ep

LP

A

MM

SM

Figure 14–4. Photomicrographs of small and large intestine and appendix. **A:** Duodenum. **B:** Large intestine. (Ep = epithelium; LP = lamina propria; MM = muscularis mucosae; A = adipocyte; SM = smooth muscle; L = lumen; GC = goblet cell; LC = lymphoid cells.) (continued)

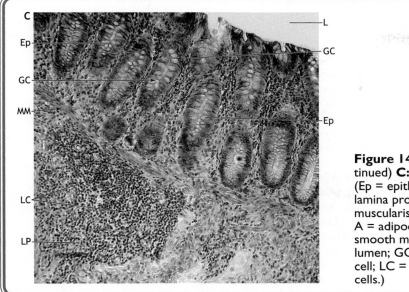

Figure 14–4. (continued) **C:** Appendix. (Ep = epithelium; LP = lamina propria; MM = muscularis mucosae; A = adipocyte; SM = smooth muscle; L = lumen; GC = goblet cell; LC = lymphoid cells.)

2. **Pancreatic lipase** promotes the hydrolysis of lipids to monoglycerides and fatty acids in the lumen.
3. **Bile acids** stabilize the hydrolytic products in an emulsion.
4. These products pass through the microvilli membrane passively and are collected in smooth endoplasmic reticulum (SER) cisternae, where they are resynthesized into **triglycerides.**
5. Triglycerides are surrounded by a layer of proteins to form **chylomicrons,** which are transferred to the **Golgi apparatus.**
6. From the Golgi apparatus, chylomicrons migrate to the lateral cell membrane, cross it by exocytosis, flow into the extracellular space, and are absorbed by **lacteals.**

D. **Components of the Wall of the Small Intestine**
1. Epithelium
   a. The **simple columnar epithelium** is replaced every **3–6 days** and is covered by a glycocalyx, consisting of glycoproteins. The absorptive cells have microvilli with bound dipeptidases and disaccharidases and contain pinocytotic vesicles.
   b. **Goblet cells** are scattered between columnar cells. They are few in number in the duodenum but gradually increase in number in the ileum. These cells produce a mucus, which protects and lubricates the intestinal lining.
   c. **Paneth's cells** are protein-secreting cells found at the base of intestinal glands. They contain secretory granules in their apical cytoplasm. **Lysozyme,** a secreted product, has a **bactericidal** effect and thus may control intestinal flora.
2. The **lamina propria** consists of loose connective tissue with nerve fibers, smooth muscle, and blood and lymphatic vessels. It contains macrophages and

lymphoid cells, which are an immunologic barrier. The lamina propria in the villi contains blood and lymphatic vessels, nerves, connective tissue, and smooth muscle.

3. Submucosa
   a. The **submucosa** in the duodenum contains coiled tubular glands, called **Brunner's glands,** which are exocrine glands found in the submucosa.
   b. These glands have ducts that penetrate the muscularis mucosa and secrete into the base of intestinal glands.
   c. These glands produce a clear, viscous **alkaline** solution that protects the small intestine from the acidic environment created by gastric juices and increases the intestinal contents to the optimum pH for action of pancreatic enzymes.
   d. The **submucosal plexus of Meissner** is found within the submucosa, whereas the **myenteric plexus of Auerbach** is found between the outer longitudinal and inner circular layers of the muscularis externa.
4. Intestinal glands (crypts of Lieberkühn).
   a. **Intestinal glands** are invaginations of surface epithelium between intestinal villi that continue to the muscularis mucosa.
   b. The upper half of these glands contains absorptive and goblet cells, whereas the lower half has numerous mitotic (stem) cells.
   c. **Paneth's cells** are found at the base of these glands, and **enteroendocrine cells** are present in the lower region.

DIGESTIVE TRACT TUMORS

- *About 90–95% of **malignant tumors** of the digestive tract are derived from intestinal or gastric epithelial cells.*

E. Peyer's Patches
1. **Peyer's patches** are found in the ileum and consist of lymphoid tissue. Each patch consists of 10–200 lymphoid nodules.
2. These clusters of lymphoid nodules, about 30 in humans, are located in the lamina propria and submucosa.
3. Peyer's patches are not covered by villi but rather by membranous epithelial cells called **M cells.**
4. **M cells** initiate **immune responses** by ingesting antigens from the intestinal lumen and transporting them to the underlying lymph nodules. These M cells are an important link within the immune system of the intestine.

VII. Large Intestine
A. **General Features**
1. The **large intestine** is ~6 feet long and has a mucosa that has no folds except in the rectum (Figure 14–4B).
2. **Water** is absorbed passively by mucosa of the colon; thus, food enters as a semifluid and exits as a semisolid.
3. Cells of the large intestine do not produce digestive enzymes, although bacteria that may perform limited digestive functions are present.
B. **Components of the Wall of the Large Intestine**
1. Epithelium
   a. The absorptive cells are **columnar** and have short microvilli.

b. Epithelial cells of the large intestine are replaced about every **6 days** by the proliferation and differentiation of **stem cells** at the base of intestinal glands.

       c. The large intestine has more **goblet cells** than the small intestine but lacks **villi** and **Paneth's cells.**

   2. The lamina propria has abundant lymphoid cells, and nodules extend into the submucosa.

   3. The muscularis consists of longitudinal and circular strands. Fibers of the outer longitudinal layer of the large intestine form **3 thick longitudinal bands** called **teniae coli.**

C. Appendix

   1. The **appendix** is a blind evagination that extends from the cecum (Figure 14–4C).

   2. It has extensive diffuse and nodular **lymphoid tissue** that forms an almost continuous layer.

   3. The appendix has no villi and few goblet cells, Paneth's cells, and intestinal glands.

   4. The appendix has more **enteroendocrine cells** than the rest of the gastrointestinal tract.

APPENDICITIS

- *Acute appendicitis is an inflammation caused by a bacterial infection that may occlude the lumen of the appendix. Pain usually is felt in the lower right quadrant of the abdomen.*

- *Chronic appendicitis is characterized by scarring or other deformity of the appendix resulting from acute appendicitis.*

D. Rectum

   1. The **rectum,** which is a continuation of the sigmoid colon, is 5 inches long and structurally similar to the large intestine.

   2. Its mucosa consists of a simple columnar epithelium with goblet cells.

   3. The rectum has a few shallow **crypts of Lieberkühn.**

E. Anal Canal and Anus

   1. The **anal canal** extends from the **anorectal junction** to the anus. **Hemorrhoids** are varicose veins found in the submucosa at the anorectal junction.

   2. The mucosa projects **longitudinal anal columns** that meet distally to form **anal valves.** Goblet cells and absorptive cells are found in the mucosa proximal to the anal valves.

   3. The epithelium of the anal canal is simple cuboidal to the **pectinate line,** which is found immediately distal to anal valves. This epithelium transitions into stratified squamous, nonkeratinized epithelium to the external anal orifice.

   4. The **anus** is lined by stratified, keratinized squamous epithelium.

TUMORS OF LARGE INTESTINE

- *Tumors of the large intestine arise primarily from its glandular epithelium, termed **adenocarcinoma.** These tumors are the second most common cause of cancer deaths in the United States.*

CLINICAL PROBLEMS

1. Which of the following is the primary function of lacteals in the small intestine?
   A. Absorption of glucose
   B. Absorption of amino acids
   C. Absorption of chylomicrons
   D. Absorption of salts
   E. Absorption of water

2. Which of the following would result from a reduction in the number of Paneth's cells?
   A. Increased levels of intestinal fats
   B. Reduced breakdown of sugars
   C. Elevated levels of undigested proteins
   D. Decreased mucus in the intestine
   E. Increased number of intestinal bacteria

3. Which of the following substances is secreted by parietal cells?
   A. Gastrin
   B. Lipase
   C. Mucus
   D. HCl
   E. Alkaline solution

4. Which of the following regions of the gastrointestinal tract has 3 layers of smooth muscle arranged in longitudinal bands?
   A. Large intestine
   B. Appendix
   C. Small intestine
   D. Stomach
   E. Esophagus

5. Which of the following is a secretion of chief cells of the fundic glands of the stomach?
   A. HCl
   B. Pepsinogen
   C. Mucus
   D. Secretin
   E. Histamine

6. In which of the following regions of the oral cavity would taste buds be located in highest concentrations?

A. Fungiform papillae

B. Filiform papillae

C. Epiglottis

D. Ventral surface of tongue

E. Circumvallate papillae

7. Replacement of epithelial cells that line the small intestine is essential because of the acidic chyme. How often are the epithelial cells replaced?

A. Every hour

B. Every 12 h

C. Every 24 h

D. Every 3–6 days

E. Every 2–3 weeks

8. Absorption of lipids occurs within the small intestine. Which of the following organelles is most essential for lipid absorption?

A. Golgi apparatus

B. Mitochondria

C. Rough endoplasmic reticulum

D. Lysosomes

E. Peroxisomes

9. Which of the following cells in the epithelium of the stomach has apical secretory canaliculi that extend into the basal region of the cell?

A. Chief cells

B. Parietal cells

C. Mucous neck cells

D. Enteroendocrine cells

E. Stem cells

10. Which of the following cell types is most prevalent within the gastric pit of the stomach?

A. Columnar absorptive cells

B. Enteroendocrine cells

C. Mucus-secreting cells

D. Stem cells

E. Paneth's cells

ANSWERS

1. The answer is C. Lacteals are found within the villi that project into the lumen of the small intestine. These lymphatic vessels absorb chylomicrons, which are formed within cells lining the small intestine.

2. The answer is E. Paneth's cells secrete lysozyme, which attacks and destroys bacteria within the intestinal tract.

3. The answer is D. Parietal cells secrete HCl into the intestinal tract, creating the acidic environment necessary for enzyme activity.

4. The answer is A. The smooth muscle layers in the large intestine from the cecum to the rectum are arranged in longitudinal bands.

5. The answer is B. Chief cells within fundic glands of the stomach secrete proenzymes, such as pepsinogen. HCl is secreted by parietal cells, whereas secretin is a hormone secreted by enteroendocrine cells.

6. The answer is E. Taste buds are most abundant on the lateral surface of circumvallate papillae. However, a few taste buds are found on fungiform papillae, mucosa of the mouth, epiglottis, and the soft palate.

7. The answer is D. The epithelial cells that line the small intestine are replaced every 3–6 days.

8. The answer is A. Fatty acids and monoglycerides diffuse across the plasma membrane of intestinal epithelia and are transformed into triglycerides within SER. These packaged triglycerides are transformed into chylomicrons and transferred to the Golgi apparatus. The vesicles migrate to the lateral membrane of the intestinal cells and are exocytosed into the extracellular space.

9. The answer is B. The parietal cell of the stomach is distinctive by the presence of extensive secretory canaliculi found at the apex of the cell. This complex membrane infolding is called a tubulovesicular system. The unique membrane allows for secretion of hydrochloric acid into the gastric juice.

10. The answer is C. The most abundant cells of a gastric pit within the stomach are mucus-secreting cells. The mucus protects the mucosa from the acidic and hydrolytic chyme.

CHAPTER 15

LIVER, GALLBLADDER, AND EXOCRINE PANCREAS

I. Liver

A. Functions

1. **Exocrine functions** of the liver include the synthesis and secretion of **bile** via a duct system.
2. **Endocrine functions** of the liver include the synthesis and secretion of plasma proteins, cholesterol, lipoproteins, and glucose directly into the blood stream.
3. **General functions** of the liver include detoxification of lipid-soluble drugs, breakdown of steroid hormones, storage of glycogen, and production of urea.

B. General Features

1. The liver is encapsulated by **Glisson's capsule.** Hepatocytes constitute the bulk of the liver, whereas connective tissue is a minor component (Figure 15–1A and B).
2. The hepatocytes are radially arranged in plates around small venules, called **central veins,** to form **classic lobules** (Figure 15–1C).
3. The **classic lobules** are hexagonal cylinders 2 mm long and 1 mm wide.
4. A **portal triad** consists of a branch of a hepatic **artery,** a branch of a portal **vein** (interlobular vein), and a branch of a bile **duct** (interlobular bile duct) (Figure 15–1D).

C. Classic Lobule

1. The **classic lobule** (Figure 15–1D and E) emphasizes the **endocrine** function of the liver; the central vein is at the center of each lobule with 6 portal triads.
2. These lobules explain how damage to the liver occurs in disease and toxic conditions by changing the constituents of blood as it flows through the lobule.

D. Portal Lobule

1. The portal lobule emphasizes the **exocrine function** of liver; the interlobular bile duct is at the center of each triangular lobule with 3 central veins at the apices (Figure 15–1E).
2. **Bile** flows from the **bile canaliculi** to terminal ductules, the interlobular bile duct, and the right and left hepatic ducts that join the cystic duct to form the common bile duct.
3. **Bile canaliculi** are formed between hepatocytes by tight junctions.

E. Liver Acinus

1. The **liver acinus,** also called **Rappaport lobule,** emphasizes metabolic gradients and the functional activity of the liver (Figure 15–1E).

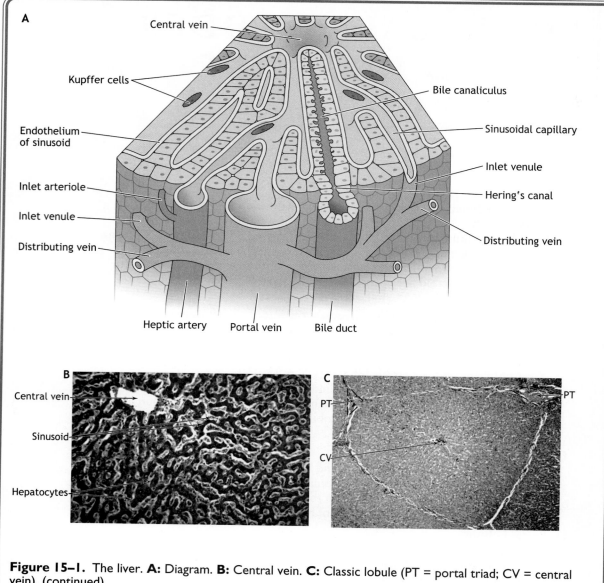

Figure 15–1. The liver. **A:** Diagram. **B:** Central vein. **C:** Classic lobule (PT = portal triad; CV = central vein). (continued)

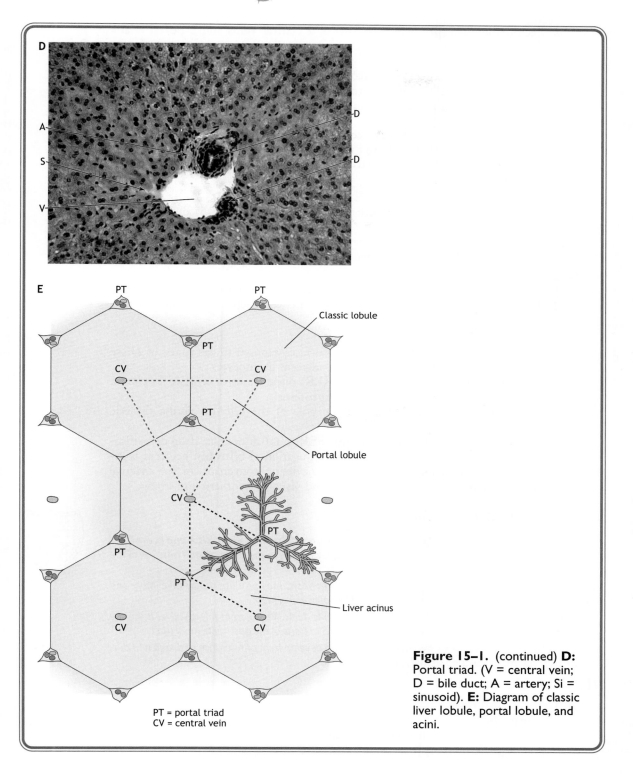

Figure 15–1. (continued) **D:** Portal triad. (V = central vein; D = bile duct; A = artery; Si = sinusoid). **E:** Diagram of classic liver lobule, portal lobule, and acini.

2. Each diamond-shaped acinus consists of **2 central veins** and **2 portal triads** that form the corner of each acinus.
3. Zones of acini
   a. **Zone I** in the acinar center receives blood first from terminal distributing vessels. These cells are the last to die and the first to regenerate.
   b. Zone I has higher oxygen, nutrients, and toxin concentrations than the other 2 zones.
   c. **Glycogen** is first deposited in **zone I** after feeding and is the zone of greatest **mitotic** activity after insult to the liver.
   d. **Zones II** and **III** are less active and have abundant lipid and aging pigments, especially in **zone III.**

F. **Sinusoids of Liver**
   1. Blood vessels between cords of hepatocytes are lined by a discontinuous endothelial layer.
   2. **Kupffer's cells** contribute to the endothelial lining or rest on it in the lumen of the sinusoid. These cells are **phagocytic,** especially for aged erythrocytes.

G. **Space of Disse**
   1. The **space of Disse** is an extracellular space between sinusoids and hepatocytes.
   2. This space is the site of traffic of materials between hepatocytes and lumen of sinusoids.

H. **Hepatocytes**
   1. **Hepatocytes** are polyhedral cells with surfaces exposed to the **space of Disse** and **bile canaliculi** and in contact with adjacent hepatocytes.
   2. The **rough endoplasmic reticulum (RER)** of hepatocytes synthesizes plasma proteins and the protein portion of **lipoproteins.**
   3. The **Golgi apparatus** is found at the pole of the cell that forms the bile canaliculus.
   4. The **smooth endoplasmic reticulum (SER)** has 6 functions: drug detoxification, breakdown of steroid hormones, conjugation of **bilirubin** before it is excreted into bile, cholesterol synthesis, enzymatic conversion of the inactive **glycogen synthase** into its active form, and glycogen synthesis and storage.

LIVER INFLAMMATORY DISORDERS

- *Inflammatory disorders, particularly those induced by viral agents such as **hepatitis A and B,** are common and can produce devastating acute and progressive damage to the structure and function of the liver.*
- *Degenerative, metabolic, and toxic responses, such as **cirrhosis,** may produce progressive injury and death of hepatocytes with subsequent scarring and loss of liver function.*
- ***Neoplasia** may develop primarily from hepatocytes or bile ducts. More commonly, the liver is the site of **secondary** or **metastatic tumors** transported from other parts of the gastrointestinal tract.*
- *The **SER** of hepatocytes becomes **hypertrophied** during development of tolerance to **drugs** and during detoxification of circulating steroids and barbiturates.*

II. Gallbladder

A. **General Features**
   1. The **gallbladder** is a diverticulum of the common hepatic duct. It is 3 in long and has a diameter of 1.5 in but can undergo considerable distention (Figure 15–2A).

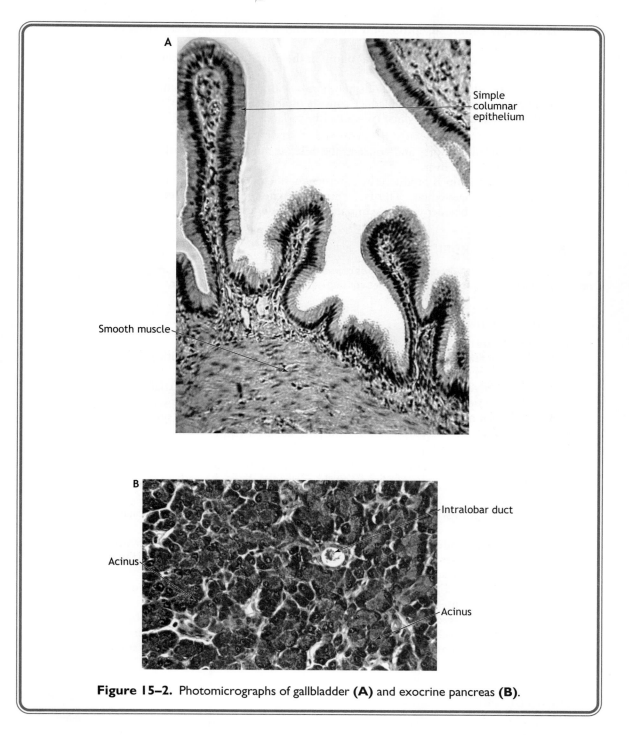

Figure 15–2. Photomicrographs of gallbladder **(A)** and exocrine pancreas **(B)**.

2. A **simple columnar epithelium** lines the lumen of the gallbladder, which extends apical microvilli.
3. These epithelial cells have an **Na⁺** pump in their basal surface to facilitate water absorption from stored **bile.**
4. The **muscularis** is composed of smooth muscle underlying the lamina propria.
5. The outer layer of the gallbladder consists of an **adventitia**, which attaches to the liver, and a serosa, which covers its free surface.

B. Function
1. The gallbladder stores and concentrates **bile,** as much as 30–50 mL, which is produced by the liver.
2. **Cholecystokinin** is produced by enteroendocrine cells of the small intestine in response to ingested fats. This factor stimulates the contraction of smooth muscle of the gallbladder to release **bile** into the **duodenum.**

GALLBLADDER DISORDER

- **Cholelithiasis** is a disorder in which **concretions** of primarily cholesterol accumulate in the gallbladder. These concretions can block the cystic duct, causing extreme pain called **biliary colic.**

III. Exocrine Pancreas

A. General Features
1. The **pancreas** is a lobulated, compound **tubuloalveolar gland** with both endocrine and exocrine functions (Figure 15–2B).
2. **Acini** or **alveoli** of the pancreas are tube shaped, surrounded by a basal lamina, and composed of 5–8 pyramidal cells arranged around a central lumen.
3. The lumen may have **centroacinar cells** that are cells of the ductal system.

B. Duct System
1. The **intercalated ducts** in the pancreas are lined by a simple cuboidal epithelium and extend into the lumen of the alveolus as **centroacinar cells.**
2. Interlobular ducts merge to form a **main excretory duct.**

C. Secretions
1. **Acini** are groups of serous cells that secrete their product into intercalated ducts.
2. **Zymogen granules** in the acinar cells are rich in digestive **enzymes,** such as lipase, trypsinogen, chymotrypsinogen, ribonucleases, and amylase. These **proenzymes** are activated within the duodenum.
3. Acinar cells are stimulated to secrete their products by **cholecystokinin,** which is released by enteroendocrine cells of the small intestine.
4. Cells of the **intercalated ducts** are stimulated to secrete their products by **secretin,** a hormone that is secreted by cells of the small intestine. This secretion is rich in **bicarbonate** and assists in neutralizing acidic chyme from the stomach.

PANCREAS DISORDER

- **Acute pancreatitis** results from viruses, drugs, and alcohol activating **proforms** of digestive enzymes such as trypsin, chymotrypsin, and other proteolytic enzymes. After activation, these enzymes actively digest cellular components of the pancreas, resulting in abdominal pain and nausea.

CLINICAL PROBLEMS

A biopsy of the gallbladder was performed on a patient suffering from abdominal pain. The mucosal lining of the lumen of the gallbladder was analyzed microscopically and determined to be normal.

1. Which of the following types of epithelia was observed?
   A. Simple squamous epithelium
   B. Transitional epithelium
   C. Pseudostratified columnar epithelium
   D. Simple columnar epithelium
   E. Stratified squamous epithelium

2. Which of the following organelles within the hepatocyte functions to detoxify drugs?
   A. Rough endoplasmic reticulum
   B. Smooth endoplasmic reticulum
   C. Golgi apparatus
   D. Lysosome
   E. Peroxisome

3. Which of the following directly stimulates ductal cells of the pancreas to secrete bicarbonate solution?
   A. Secretin
   B. Cholecystokinin
   C. High stomach acid
   D. Gastrin
   E. Postganglionic nerves

4. Which of the following form the space of Disse within the liver?
   A. Adjacent hepatocytes
   B. Adjacent endothelia of liver sinuses
   C. Between venules, arterioles, and lymphatics of a portal triad
   D. Kupffer's cells and sinus endothelium
   E. Sinus endothelium and hepatocytes

5. Which of the following are components of a portal triad within the liver?
   A. Bile canaliculi and lymphatic vessels
   B. Three central veins
   C. Central vein and hepatic sinuses
   D. Hepatic artery, portal vein, and bile duct
   E. Common bile, hepatic bile, and cystic ducts

6. The contents of zymogen granules within acinar cells of the pancreas are secreted into the intercalated duct. Which of the following provide the primary stimulation for this secretion?

    A. Bicarbonate

    B. Secretin

    C. Cholecystokinin

    D. Acid chyme

    E. Digestive enzymes

A pathologist is examining liver tissue sections. He notes large accumulations of glycogen within hepatocytes. He knows glycogen is synthesized and stored within the liver.

7. Which of the following organelles functions to synthesize glycogen?

    A. Rough endoplasmic reticulum

    B. Smooth endoplasmic reticulum

    C. Golgi apparatus

    D. Polyribosomes

    E. Lysosomes

8. Within the liver, blood within the sinusoids drains into which of the following?

    A. Central vein

    B. Portal venule

    C. Bile canaliculi

    D. Hepatic arteriole

    E. Portal lobule

9. Which of the following defines the classic liver lobule?

    A. Triangular with a central vein and 3 portal triads

    B. Triangular with a central portal triad and 3 central veins

    C. Quadrangular with 2 central veins and 2 portal triads

    D. Hexagonal with a central portal triad and 6 central veins

    E. Hexagonal with a central vein and 6 portal triads

10. Bile formed within the liver is transported via canaliculi to bile ducts. Which of the following form the canaliculi?

    A. Central veins

    B. Endothelia

    C. Kupffer's cells

    D. Hepatocytes

    E. Merging sinusoids

ANSWERS

1. The answer is D. The lumen of the gallbladder is lined with a simple columnar epithelium.

2. The answer is B. Smooth endoplasmic reticulum of hepatocytes has several functions, including detoxification of drugs.

3. The answer is A. Ductal cells in the pancreas secrete a bicarbonate solution in response to secretin, a hormone secreted by enteroendocrine cells of the small intestine.

4. The answer is E. The space of Disse is bound by the basal region of hepatocytes and the endothelium of hepatic sinuses. Macrophages are found attached to the endothelium positioned to phagocytize damaged erythrocytes and foreign debris.

5. The answer is D. A portal triad consists of a branch of a hepatic artery, the portal vein, and a bile duct.

6. The answer is C. The hormone cholecystokinin stimulates the release of the contents of zymogen granules after binding to receptors on the surface of hepatocytes.

7. The answer is B. The smooth endoplasmic reticulum is the site of glycogen synthesis. Glycogen can also accumulate within the SER.

8. The answer is A. Liver sinusoids drain into central veins that are at the center of a liver lobule. Portal veins and hepatic arterioles drain into sinusoids.

9. The answer is E. A classic liver lobule is demarcated as a hexagon with a single central vein and 6 portal triads at the periphery. A portal lobule is delineated as a triangle with a single central portal triad and 3 central veins at the apices. A liver acinus is quadrangular with 2 portal triads and 2 central veins.

10. The answer is D. Bile canaliculi are formed by the space created between 2 adjacent hepatocytes. This small canal is maintained by tight junctions.

CHAPTER 16
RESPIRATORY SYSTEM

I. Components of Respiratory System

A. Conducting Portion

1. The **conducting portion** of the respiratory system consists of the nasal cavity, nasopharynx, larynx, trachea, bronchi, bronchioles, and terminal bronchioles (Figure 16–1).

2. This portion provides a channel through which air can pass to and from the lungs, and it also **conditions** the air. Before entering the lungs, air is cleansed, moistened, and warmed by the **respiratory epithelium,** which contains mucous and serous glands and a vascular network.

B. Respiratory Portion

1. The **respiratory portion** consists of respiratory bronchioles, alveolar ducts, and alveoli.

2. The exchange of oxygen and carbon dioxide occurs within this portion.

II. Respiratory Epithelium

A. Pseudostratified ciliated columnar epithelium (PCCE) lines most of the conducting portion of the respiratory system (Figure 16–2A).

B. The PCCE gives way to a simple columnar epithelium and then to a simple cuboidal epithelium in the terminal bronchioles.

C. Cells of respiratory epithelium include ciliated columnar cells, goblet cells, brush cells, basal cells, and granule cells.

1. **Ciliated columnar cells** are the most abundant cells. Each cell possesses about 300 cilia, which require adenosine triphosphate (ATP) for their movement.

2. **Goblet cells** contain mucous droplets in their apical cytoplasm.

3. **Brush cells** are columnar in shape and have numerous microvilli on their apical surface.

4. **Basal cells** are small cells that lie on the basal lamina but do not reach the luminal surface of the epithelium. These cells may be mitotically active and differentiate into respiratory cell types.

5. Small **granule cells** have 100–300-nm granules within their cytoplasm. These cells are part of the **diffuse neuroendocrine system** (DNS), which contains molecules such as polypeptide hormones and amines, including epinephrine, serotonin, and norepinephrine.

D. The **goblet cell** population in the epithelium gradually reduces, and these cells are absent in terminal bronchioles. **Ciliated cells** are present where the goblet

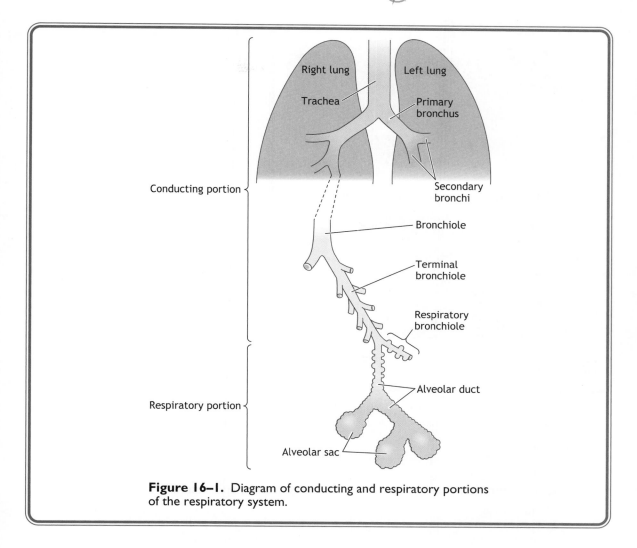

Figure 16–1. Diagram of conducting and respiratory portions of the respiratory system.

cells are lacking in the finer or deeper bronchioles to prevent mucus accumulation in the respiratory portion of the system.

E. A secretion of **serous glands** floats on the cilia of the epithelia. These cilia then move this fluid, together with the overlying mucus and trapped particles, toward the oral cavity, where it is either swallowed or expectorated.

LUNG TUMORS

- *Tumors of the lung are mainly **epithelial** in origin. **Squamous cell carcinoma,** the principal lung cancer, is caused by smoking, which affects the epithelium of the bronchi and bronchioles.*
- *Chronic smoking induces the **transformation** of the respiratory epithelium into a stratified squamous epithelium, which is an initial step in its eventual differentiation into a tumor.*

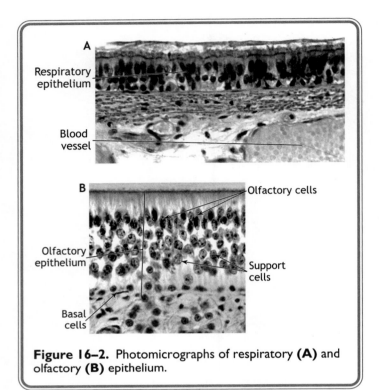

Figure 16–2. Photomicrographs of respiratory **(A)** and olfactory **(B)** epithelium.

III. Nasal Cavity

A. **Vestibule**

1. The **vestibule,** the most anterior and dilated part of the nasal cavity, contains the nares and vibrissae.
2. The epithelium changes from keratinized to respiratory epithelium.

B. **Nasal Fossae**

1. **Conchae** are bony, shelflike projections from the lateral wall within the nasal cavity. The middle and inferior projections are lined with respiratory epithelium, whereas the superior conchae are lined by olfactory epithelium.
2. **Olfactory epithelium** is a pseudostratified columnar epithelium consisting of 3 cell types (Figure 16–2B).
   a. **Supporting cells** have a narrow base but a broad cylindrical apex. Microvilli are seen at its surface, which is bathed with a fluid consisting of serous and mucous secretions.
   b. **Basal cells** are small spherical or cone-shaped cells that form a single layer at the base of the epithelium.
   c. **Olfactory cells** are **bipolar neurons** found between basal and supporting cells. Their nuclei lie below those of supporting cells. These cells have 6–20 **cilia,** which are long and nonmotile and possess the receptors for odoriferous substances. The afferent axons of the bipolar neurons form small bundles toward the central nervous system (CNS).

SWELL BODIES

- *Swell bodies* are found in the lamina propria of the conchae. These structures are large venous plexuses, which on 1 side of the nasal fossae become engorged with blood every 20–30 min. This results in a distention of the **conchae mucosa** and decrease in airflow. During this time, most of the air is directed through the opposite nasal fossa, which then allows the respiratory epithelium to recover from desiccation on the side with reduced airflow.

IV. Paranasal Sinuses

A. The **paranasal sinuses** are blind cavities in the frontal, ethmoid, maxillary, and sphenoid bones, which are lined by a thin respiratory epithelium.

B. These sinuses communicate with the nasal cavity through small openings. Mucus formed in these cavities drains into the nasal passages through these openings by the action of ciliated epithelial cells.

SINUSITIS

- *Sinusitis* is an inflammatory process that may persist as a result of blockage of the drainage orifices.
- *Chronic sinusitis* is a component of **Kartagener's syndrome,** which results from defective ciliary action.

V. Nasopharynx

A. The **nasopharynx** is the first part of the pharynx and continues caudally with the oropharynx.

B. This region is lined by respiratory epithelium in the portion that is in contact with the soft palate.

VI. Larynx

A. The **larynx** connects the pharynx with the trachea.

B. Large **hyaline cartilages** and smaller **elastic cartilages** in its lamina propria are bound together by ligaments (Figure 16–3A).

C. The **epiglottis** projects from the rim of the larynx and extends into the pharynx. The lingual and apical surfaces of the laryngeal side are lined by stratified squamous epithelium. Toward the base of the epiglottis on the laryngeal surface, the epithelium changes to a respiratory epithelium.

D. Two pairs of folds, below the epiglottis, extend into the lumen of the larynx. The upper pair is the **false vocal cord** or **vestibular fold** and is lined by respiratory epithelium.

E. The lower pair of folds is the **true vocal cord,** which is lined with nonkeratinized stratified squamous epithelium. The vocal ligament, large bundles of elastic fibers, lies within this fold. Bundles of skeletal muscle, **vocalis muscle,** lie parallel to the ligaments and regulate the tension of the fold and ligament.

VII. Trachea

A. The **trachea** (Figure 16–3B) is a 10-cm long tube that extends from the larynx to a point where it bifurcates into the 2 primary bronchi.

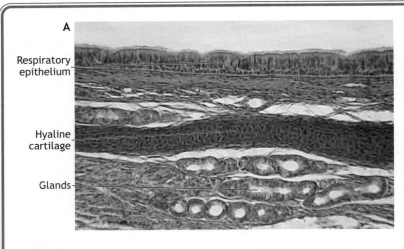

A

Respiratory epithelium

Hyaline cartilage

Glands

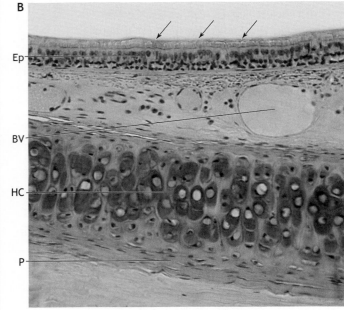

B

Ep

BV

HC

P

Figure 16–3. Photomicrographs of larynx **(A)** and trachea **(B).** (Ep = respiratory epithelium; BV = blood vessel; HC = hyaline cartilage; P = perichondrium. Arrows point to cilia.)

B. The trachea, lined by respiratory epithelium, has a lamina propria that contains 16–20 **C-shaped rings of hyaline cartilage.**

C. The **fibroelastic ligament** and **trachealis muscle** bridge the opening of the tracheal cartilage.

D. Contraction of the muscle and resultant narrowing of the tracheal lumen, or **cough reflex,** leads to increased expired air, which aids in clearing the air passage.

VIII. Bronchial Tree

A. General Features

1. The trachea divides into 2 **primary bronchi** that enter the lung at the hilus. Arteries enter, and veins and lymphatic vessels leave the lung at the hilus (Figure 16–1).
2. The structures in this region are surrounded by the **root of the lung.**

B. Bronchi

1. The primary bronchi within the lung divide into **secondary bronchi**: 3 in the right lung and 2 in the left lung.
2. The primary bronchi are very similar histologically to the trachea.
3. Each primary bronchus divides 9–12 times, decreasing in size until the branches become about 5 mm in diameter.
4. The cartilage within the lamina propria of secondary bronchi appears as **irregular plates** or **islands.**
5. With further division, **smooth muscle** becomes more abundant in the lamina propria with elastic fibers.
6. Ducts of mucous and serous glands open into the bronchial lumen.

ASTHMA

- *Asthma is a condition in which the passageways within the lungs become narrow, usually in the bronchi.*
- *This condition may result from edema of the mucosa or spasm of smooth muscle of bronchi. Another cause may be release of **histamine** or **prostaglandins** after an allergic response.*
- *The drug **isoproterenol** is used to effect bronchial dilatation during **asthma attacks.***

C. Bronchioles

1. **Bronchioles** (Figure 16–4A) are 5 mm or smaller in diameter and lack glands and cartilage, with only a few goblet cells.
2. **Larger bronchioles** have a **ciliated columnar epithelium,** whereas smaller terminal bronchioles are lined by a ciliated cuboidal or simple cuboidal epithelium, called terminal bronchioles.
3. **Terminal bronchioles** have **Clara cells,** which secrete glycosaminoglycans. These cells are dome shaped with no cilia.
4. The lamina propria of bronchioles has abundant smooth muscle and elastic fibers. The muscle is controlled by the vagus nerve and sympathetic nervous system. Stimulation by the **vagus nerve** decreases the size of the bronchiole lumen, whereas **sympathetic** stimulation has the opposite effect.

D. Respiratory Bronchioles

1. Each terminal bronchiole divides into 2 or 3 respiratory bronchioles.
2. **Respiratory bronchioles** (Figure 16–4B) are the first segments of the respiratory portion of the respiratory system, as **alveoli,** regions of gas exchange, are present in their walls.
3. These bronchioles are lined with a ciliated cuboidal epithelium and **Clara cells.**
4. At the rim of the alveolar openings, the epithelium becomes continuous with the **type I alveolar cells.**
5. The number of alveoli increases in more distal respiratory bronchioles.

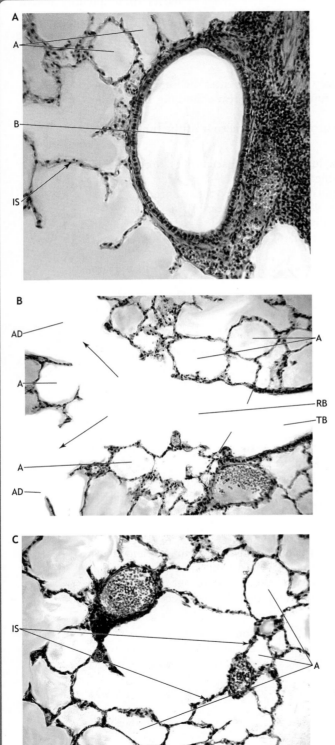

Figure 16–4. Photomicrographs of bronchiole **(A),** terminal bronchiole (arrows show direction of air flow) **(B),** and alveoli **(C).** (B = bronchiole; A = alveolus; IS = interalveolar septum; AD = alveolar duct; TB = terminal bronchiole; RB = respiratory bronchiole; BV = blood vessel.)

6. Smooth muscle and elastic fibers lie beneath the epithelium of respiratory epithelium.

E. Alveolar Ducts

1. The **alveolar ducts** are regions of the bronchiole in which only alveolar openings are found.
2. Alveolar ducts and alveoli are lined by **type I alveolar cells,** and the lamina propria contains smooth muscle.
3. The alveolar ducts open into an atrium that communicates with an alveolar sac.
4. **Elastic fibers** enable the alveoli to expand during inspiration and contract during expiration.
5. **Reticular fibers** prevent overdistention of the alveoli and damage to the thin capillaries.

F. Alveoli

1. **Alveoli** (Figure 16–4C) are saclike structures (about 200 μm in diameter) within the parenchyma of the lung and number about 300 million.
2. Oxygen and carbon dioxide exchange with blood and air occurs within these structures.
3. An **interalveolar wall** or septum lies between adjacent alveoli and consists of 2 thin squamous epithelial layers between which lie capillaries, fibroblasts, elastic and reticular fibers, and macrophages.
4. The **blood–air barrier** consists of surfactant, cytoplasm of type I alveolar cells, basal laminae of type I alveolar cells, basal lamina of capillary endothelial cells, and the cytoplasm of endothelial cells.
5. The total thickness is 0.1–1.5 μm. A basement membrane is formed by the fusion of the basal laminae of type I alveolar cells and capillary endothelial cells.

ALVEOLAR DAMAGE

• *Following damage to the **alveoli**, the **type II alveolar cells** undergo mitotic division. These cells undergo a transformation into **type I alveolar cells** and replace those cells that were lost.*

G. Components of Interalveolar Septum

1. The **interalveolar septum** is the thinnest barrier between the blood and air and contains types I and III collagen.
2. **Endothelial cells** of capillaries are continuous and nonfenestrated. These cells have numerous pinocytotic vesicles.
3. **Type I alveolar cells** cover 97% of the luminal surface of alveoli. These cells are joined by desmosomes and zonulae occludens. The main function of these cells is to provide a barrier of minimal thickness that is readily permeable to gases.
4. **Type II alveolar cells** are cuboidal cells that rest on the basement membrane and are interspersed with type I alveolar cells to which they are connected by desmosomes and **zonulae occludens.** These cells manufacture lamellar bodies that contain **surfactant,** which performs an essential pulmonary function.
5. **Alveolar macrophages,** or dust cells, are found on the surface of the interalveolar septum. These cells are derived from monocytes that originate in the bone marrow.

6. **Pores** within the interalveolar septum, 10–15 μm in diameter, connect adjacent alveoli and function to **equalize pressure** in alveoli and allow for the **collateral circulation** of air when a bronchiole is obstructed.

EMPHYSEMA

- The destruction of the interalveolar septum with resulting reduction in the respiratory portion of the lungs is called **emphysema.**
- This disease is typically a result of environmental air pollution and smoking.

H. Surfactant
1. **Surfactant** is a layer consisting of an aqueous, protein-containing hypophase-covered monomolecular phospholipid film (dipalmitoyl lecithin).
2. Surfactant reduces **surface tension** of alveolar cells (ie, less inspiratory force is needed to inflate the alveoli, thus reducing the work of breathing). Alveoli would collapse without surfactant during expiration.

RESPIRATORY DISTRESS

- In premature infants, labored breathing indicates **respiratory distress** or hyaline membrane disease. Respiratory distress syndrome (RDS) results from insufficient surfactant production by **type II alveolar cells.**
- **Surfactant** synthesis can be stimulated by **thyroxine** and during pregnancy by glucocorticoid administration.

CLINICAL PROBLEMS

1. Which of the following types of epithelia line the true vocal fold?
   A. Keratinized, stratified squamous epithelium
   B. Nonkeratinized, stratified squamous epithelium
   C. Pseudostratified, ciliated columnar epithelium
   D. Simple cuboidal epithelium
   E. Simple columnar epithelium

2. Which of the following histologic characteristics distinguishes a bronchus within the lung from primary bronchus?
   A. Glands in the submucosa
   B. Pseudostratified ciliated columnar epithelium
   C. Smooth muscle in the walls
   D. Irregular plates of hyaline cartilage
   E. Goblet cells in the mucosa

3. Which of the following is the first component of the respiratory segment of the respiratory system?

A. Respiratory bronchiole

B. Terminal bronchiole

C. Alveolar duct

D. Secondary bronchus

E. Primary bronchus

A coal miner presents with a chronic cough. A lung biopsy reveals cells with large black deposits.

4. Which of the following are the cells containing this inhaled material?

A. Type II alveolar cells

B. Type I alveolar cells

C. Ciliated columnar epithelial cells

D. Goblet cells

E. Alveolar macrophages

5. Which of the following is lined by a pseudostratified ciliated columnar epithelium?

A. Vocal fold

B. Superior nasal conchae

C. Primary bronchi

D. Respiratory bronchiole

E. Lingual surface of the epiglottis

6. Which of the following cells line more than 90% of alveolar septum?

A. Type II alveolar cells

B. Endothelial cells

C. Fibroblasts

D. Type I alveolar cells

E. Macrophages

7. Clara cells are found almost exclusively within which component of the bronchial tree?

A. Terminal bronchioles

B. Primary bronchi

C. Alveolar ducts

D. Alveoli

E. Intrapulmonary bronchi

8. Diseased lungs with a diminished ability to recoil would exhibit a defect in or loss of which of the following?

A. Reticular fibers

B. Elastic fibers

C. Smooth muscle

D. Collagen fibers

E. Type I alveolar cells

ANSWERS

1. The answer is B. The true vocal fold is lined by a nonkeratinized, stratified squamous epithelium. The epithelium gradually changes back to a respiratory epithelium inferior to the trachea.

2. The answer is D. The major difference between a bronchus within the lung, called an intrapulmonary bronchus, and an extrapulmonary bronchus is the presence of plates or islands of hyaline cartilage in its wall. The primary bronchi have C-shaped cartilage.

3. The answer is A. Respiratory bronchioles comprise the first segment of the respiratory tract that possesses alveoli within its wall. Thus, gas exchange occurs here first.

4. The answer is E. Alveolar macrophages phagocytize inhaled particles, including coal dust. These cells will remain within the lung tissues.

5. The answer is C. The lumina of the primary bronchi and trachea are lined by pseudostratified ciliated columnar epithelium.

6. The answer is D. The thin type I alveolar cell, also called a type I pneumocyte, covers 97% of the alveolar septum. Type II alveolar cells line a limited area of the septum. Macrophages are dispersed within the alveolus.

7. The answer is A. Clara cells are found lining the lumen of terminal and respiratory bronchioles. These cells secrete a glycoprotein that acts similarly to surfactant.

8. The answer is B. Elastic fibers allow the lung and alveoli to recoil after expiration. Reticular fibers prevent overextension of the lungs during inspiration.

CHAPTER 17
URINARY SYSTEM

I. Kidney Functions

   A. The kidney **filters** blood and **removes** waste products of metabolism, resulting in the production of urine.

   B. The kidney also regulates fluid and electrolyte balance, and renal cells secrete hormones, including **renin** and **erythropoietin.**

II. Kidney Structure

   A. The renal arteries and veins enter and leave the kidney at the **hilus,** which is positioned medially (Figure 17–1).

   B. The **cortex** of the kidney is encapsulated and contains as many as 4 million glomeruli.

   C. The **medulla** contains 10 or more **renal pyramids,** each of which terminates at its base as a **renal papilla** into the **minor calyx.**

   D. The minor calyces merge to form 2–3 **major calyces,** which in turn merge to form the dilated **renal pelvis.** The renal pelvis continues as the **ureter,** which drains urine into the urinary bladder.

   E. The **renal pyramids** contain parallel tubules that penetrate the cortex as **medullary rays** at the center of a **renal lobule. Medullary rays** consist of a collecting duct and the nephrons that drain into them.

III. Nephron and Uriniferous Tubules

   A. The **nephron** is the major functional unit of kidney. **Uriniferous tubules** consist of a nephron and the collecting duct that it drains.

   B. The **nephron** includes the **renal corpuscle,** consisting of a glomerulus and a Bowman's capsule, proximal convoluted tubule, thin and thick limbs of the loop of Henle, and the distal convoluted tubule.

   C. The glomerulus (Figure 17–2A) consists of a capillary bed, fenestrated endothelial cells with no diaphragms, and **mesangial cells.**

      1. Bowman's capsule is a double-walled epithelial layer that encloses the glomerulus.

      2. The external layer of Bowman's capsule is the **parietal layer** of simple squamous epithelium.

      3. The internal layer of Bowman's capsule is the **visceral layer** of **podocytes,** with primary processes and secondary processes, called **pedicles,** which attach to the capillaries of the glomerulus.

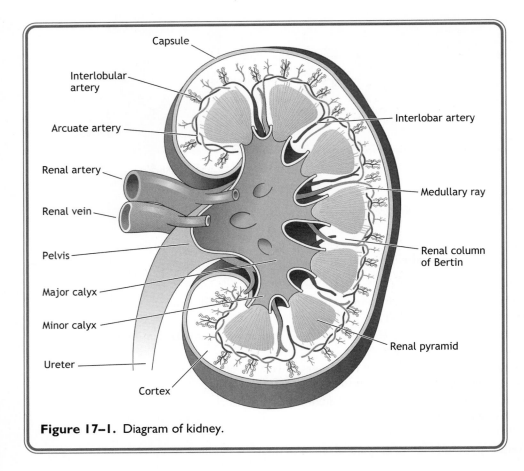

Figure 17–1. Diagram of kidney.

4. The **urinary space** is found between the parietal and visceral layers.
5. The **vascular pole** is the site of the entering afferent arteriole and the exiting efferent arteriole, whereas the **urinary pole** is the site of the origin of the proximal convoluted tubule.
6. **Filtration slits** are located between adjacent pedicles of podocytes along the **basal lamina,** which is formed by podocytes and endothelial cells. The **basal lamina** encloses the capillaries of glomeruli and intraglomerular mesangial cells (Figure 17–2B).
7. The **lamina densa** of the **basal lamina** acts as a physical filter, whereas the **lamina rara** acts as a charge barrier.

GLOMERULONEPHRITIS

- *Glomerulonephritis is an inflammation of the kidney, specifically of the capillary components of glomeruli. This condition can be chronic, acute, or rapidly progressive. **Capillaries** of the glomeruli become damaged as a result of infusion of neutrophils and lymphocytes, which leads to erythrocytes and high protein within the urine.*

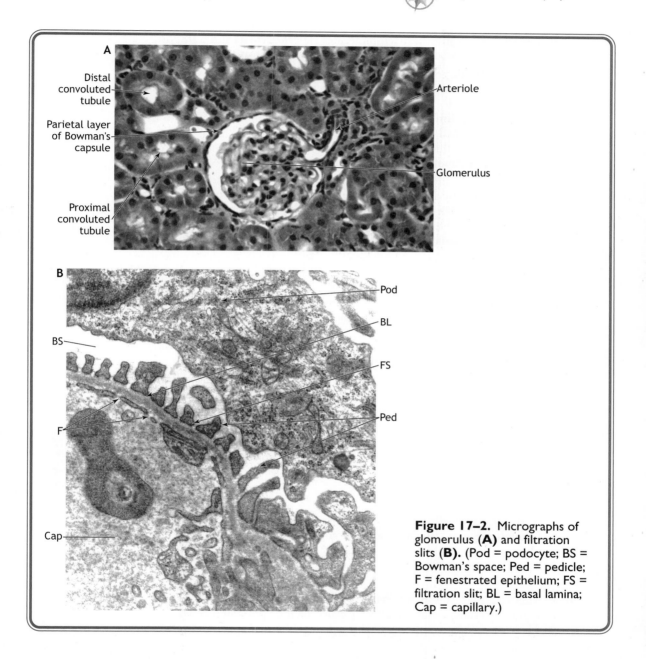

A

Distal convoluted tubule

Parietal layer of Bowman's capsule

Proximal convoluted tubule

Arteriole

Glomerulus

B

Pod

BL

BS

FS

F

Ped

Cap

Figure 17–2. Micrographs of glomerulus (**A**) and filtration slits (**B**). (Pod = podocyte; BS = Bowman's space; Ped = pedicle; F = fenestrated epithelium; FS = filtration slit; BL = basal lamina; Cap = capillary.)

GLOMERULOSCLEROSIS

• *Glomerulosclerosis is a disorder marked by **hyaline deposits** or nodules formed within the glomeruli, including the capillary basement membrane and mesangial cell.*

   D. Proximal convoluted tubules (PCT) within the **cortex** are lined by a simple cuboidal or columnar epithelium with a **brush border (microvilli).**

1. The epithelium absorbs macromolecules via **pinocytosis** and transports Na^+.
2. Because of this process, the epithelium has abundant **mitochondria.**

E. The **loop of Henle** consists of the descending limb, which is a continuation of the PCT, and the ascending limb, which continues as the distal convoluted tubule (DCT).
   1. The **ascending limb** reabsorbs filtered NaCl, K^+, and Ca^{2+}, whereas filtered water is reabsorbed by the descending loop of Henle.
   2. Each limb has a thick and thin segment. Thick segments are lined by simple cuboidal epithelium, and thin segments are lined by simple squamous epithelium.

F. The DCT is lined by a **simple cuboidal epithelium** and has no brush border.
   1. The **macula densa** is a modified portion of the **DCT** that lies adjacent to the vascular pole of the renal corpuscle.
   2. Cells of the **macula densa** monitor the Na^+ within the tubular fluid.

G. Collecting tubules, ducts, and calyces (Figure 17–3A) act as a drainage system.
   1. The DCT continues as collecting tubules that are lined by a **simple cuboidal epithelium.**
   2. **Collecting tubules** drain into collecting ducts, which in turn drain into **papillary ducts,** also called **ducts of Bellini.** The latter are lined by a simple columnar epithelium.
   3. Urine drains from the papillary ducts into **minor calyces** to **major calyces** to **renal pelvis,** which are lined by **transitional epithelium.**
   4. **Antidiuretic hormone** (ADH), also called **vasopressin,** concentrates the urine (**hypertonic**) by affecting increased permeability of water and urea by collecting tubules.

KIDNEY TUMOR

• *Mutation of both copies of the growth-regulating gene, **WT-1** (Wilms' tumor), results in a mixed tumor of the **kidney.** This tumor usually arises in children 5 years and younger.*

IV. **Juxtaglomerular Apparatus**
   A. **Juxtaglomerular Cells**
      1. **Juxtaglomerular cells** are modified smooth muscle cells of **afferent arterioles.**
      2. These cells secrete the proteolytic enzyme **renin** under stimulation by sympathetic nerve fibers.
   B. **Extraglomerular Mesangial Cells**
      1. Extraglomerular mesangial cells are called **lacis** or **polkissen cells.**
      2. These are cellular components of the juxtaglomerular apparatus.

V. **Regulation of Blood Pressure**
   A. **Renin** converts **angiotensinogen** to **angiotensin I,** which is in turn converted to **angiotensin II** by an **endothelial angiotensin-converting enzyme** (ACE).
   B. **Angiotensin II** increases blood pressure by constricting arterioles and stimulating **aldosterone** secretion from the adrenal glands.
   C. **Aldosterone** acts on cells of renal tubules to increase absorption of Na^+ and Cl^+. This expands fluid volume and increases blood pressure.

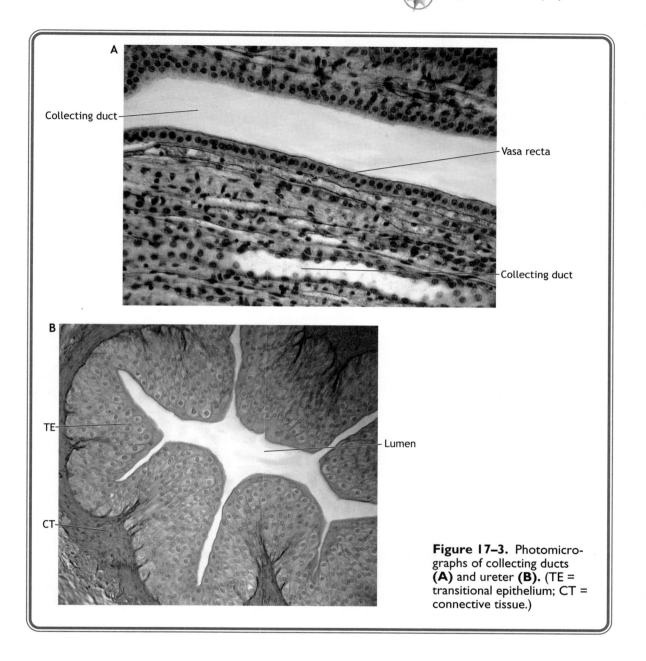

Collecting duct

Vasa recta

Collecting duct

TE

Lumen

CT

Figure 17–3. Photomicrographs of collecting ducts **(A)** and ureter **(B).** (TE = transitional epithelium; CT = connective tissue.)

VI. Blood Supply

A. Arterial Blood Supply

1. **Renal arteries,** branches of the abdominal aorta, supply the kidney and branch to form interlobar arteries.
2. **Interlobar arteries** are found between renal pyramids and branch to form **arcuate arteries,** found at the junction of cortex and medulla.

3. **Interlobular arteries** branch from arcuate arteries and traverse the cortex radially between cortical lobules.

4. **Afferent arterioles** branch from interlobular arteries and supply blood to capillaries of the glomerulus.

5. **Efferent arterioles** exit the glomerulus.

B. **Peritubular Capillaries**

1. **Peritubular capillaries** are fed by efferent arterioles of cortical nephrons, which supply proximal and distal tubules.

2. These vessels carry away absorbed ions and low-molecular-weight proteins.

C. **Vasa Recta**

1. The **vasa recta** are capillaries fed by efferent arterioles of juxtamedullary nephrons found parallel to long loops of Henle in the medulla.

2. The vasa recta supply the **medulla.**

VII. **Ureter**

A. The **ureter** extends from the **renal pelvis** to the **urinary bladder** (Figure 17–3B).

B. The lumen of the ureter is lined with a **transitional epithelium,** and its lamina propria contains smooth muscle arranged in bundles.

C. Smooth muscle contraction affects **peristalsis,** which forces urine toward the urinary bladder.

NEPHROLITHIASIS

· **Nephrolithiasis** *is the condition in which* **stones** *consisting of uric acid, calcium salts, and magnesium-ammonium acetate concentrate within the kidney. These stones can block the* **ureter,** *causing pain radiating to the side.*

VIII. **Urinary Bladder**

A. The lumen of the urinary bladder is lined by a **transitional epithelium.** All cells are interconnected by **desmosomes,** whereas cells at the luminal surface are joined by tight junctions.

B. The muscular layer consists of 3 layers of smooth muscle.

CLINICAL PROBLEMS

A pathologist is examining renal tissue with a light microscope. He notices a tubular structure that has a brush border.

1. Which of the following structures is the pathologist examining?

A. Distal convoluted tubule

B. Proximal convoluted tubule

C. Vasa recta

D. Thin loop of Henle

E. Collecting tubule

2. Which of the following cells within the glomerulus form the filtration slits?

   A. Podocytes

   B. Mesangial cells

   C. Juxtaglomerular cells

   D. Extraglomerular cells

   E. Capillary endothelial cells

3. Which of the following arteries provides blood most directly to the glomerulus?

   A. Efferent arteriole

   B. Arcuate artery

   C. Afferent arteriole

   D. Interlobular artery

   E. Interlobar artery

4. Urine passes from the Bowman's (urinary) space into which of the following structures?

   A. Collecting tubule

   B. Minor calyx

   C. Collecting duct

   D. Proximal convoluted tubule

   E. Renal pelvis

5. Which of the following is the function of the cells of the macula densa?

   A. Absorption of water after stimulation by aldosterone

   B. Secretion of renin

   C. Regulation of blood flow and glomerular filtration

   D. Absorption of water after stimulation of antidiuretic hormone

   E. Monitors Na^+ of renal tubule fluid

6. A renal lobule consists of which of the following components?

   A. Glomeruli and afferent and efferent arterioles

   B. Glomeruli and proximal and distal convoluted tubules

   C. Collecting ducts and nephrons

   D. Loop of Henle and collecting tubules

   E. Nephron and collecting duct

A large tumor is removed from the abdomen of a 5-year-old boy. Pathologic analysis reveals a Wilms' tumor.

7. From which of the following tissues would this tumor originate?

   A. Kidney

   B. Liver

C. Adrenal gland

    D. Pancreas

    E. Stomach

8. Which of the following is surrounded by the basal lamina within the kidney?

    A. Afferent and efferent arterioles

    B. Collecting tubules and ducts

    C. Proximal and distal convoluted tubules

    D. Glomerular capillaries and intraglomerular mesangial cells

    E. Vasa recta

9. Reduced water absorption within the kidney would be associated with a defect in which of the following structures?

    A. Distal convoluted tubules

    B. Collecting ducts

    C. Minor calyx

    D. Ascending and descending loops of Henle

    E. Proximal convoluted tubules

10. Blood pressure increases in response to angiotensin II by which of the following?

    A. Causing the heart to pump faster

    B. Causing constriction of arterioles

    C. Stimulating the secretion of vasopressin

    D. Increasing water uptake by cells in the collecting tubules

    E. Increasing renin secretion

You have been provided an antibody to renin. You want to identify and determine the distribution of those cells within the kidney that express this protein by immunocytochemistry.

11. Which of the following cells would you have identified?

    A. Mesangial cells

    B. Cells of proximal convoluted tubules

    C. Cells of afferent arterioles

    D. Cells of distal convoluted tubules

    E. Podocytes

12. Which of the following features is common to the calyces of the kidney, ureter, and urinary bladder?

    A. Respond to aldosterone

    B. Allow water to pass from the luminal contents

    C. Contain mucous glands in submucosa

    D. Lined by transitional epithelium

    E. Surrounded by a layer of skeletal muscle

ANSWERS

1. The answer is B. Proximal convoluted tubules are lined by a simple cuboidal epithelium. These cells extend microvilli into the lumen, which creates a brush border.

2. The answer is A. Filtration slits along the basal lamina of the glomerulus are formed by pedicles of podocytes. These cells form the visceral layer of the Bowman's capsule.

3. The answer is C. Blood to the glomerulus is supplied by the afferent arteriole, which branches from an interlobular artery. The efferent arteriole exits the glomerulus at the vascular pole.

4. The answer is D. Ultrafiltrate of blood plasma first passes through filtration slits into the Bowman's space and then into the proximal convoluted tubule.

5. The answer is E. The macula densa monitors the level of sodium ions in the filtrate that passes within the distal convoluted tubule.

6. The answer is C. A renal lobule consists of nephrons and the collecting ducts that drain them.

7. The answer is A. Wilms' tumor arises from cells and tubules of the kidney in young children.

8. The answer is D. The basal lamina surrounds the glomerular capillaries and the interposed intraglomerular mesangial cells. It serves as a barrier between the blood within the capillaries and Bowman's space.

9. The answer is E. Water is absorbed by cells of the proximal convoluted tubules.

10. The answer is B. Angiotensin II causes an increase in blood pressure by constriction of arterioles.

11. The answer is C. Renin is synthesized by specialized cells of the tunica media of afferent arterioles.

12. The answer is D. The renal minor and major calyces, ureter, and urinary bladder are all lined by transitional epithelium.

I. Pituitary Gland

A. General Features

1. The **pituitary gland** (Figure 18–1A) is found in a depression called the **sella turcica** of the sphenoid bone.

2. Hypophysis (Pituitary)
   a. The **adenohypophysis** develops from dorsal outgrowth of stomatodeum, called the pouch of Rathke, and consists of a **pars tuberalis, pars distalis,** and **pars intermedia.**
   b. **Releasing hormones** regulate the synthesis and secretion of cells in **adeno-hypophysis.** These hormones are small peptides synthesized in the **hypo-thalamus** that stimulate or inhibit secretion and are carried by a **hypophysial portal system.**
   c. The **neurohypophysis** develops from an evagination in the floor of the di-encephalon and consists of a **pars nervosa** and the **infundibulum.**

B. Blood Supply

1. **Inferior hypophysial arteries** supply blood to the pars nervosa and anterior lobe.

2. **Superior hypophysial arteries** supply a capillary bed in the median eminence of the hypothalamus.

3. **Capillaries** are collected into veins that empty into fenestrated capillaries of the adenohypophysis, which form the **hypophysial portal system.**

C. Adenohypophysis (Anterior Pituitary Gland)

1. Pars distalis (anterior lobe) (Figure 18–1B)
   a. Cords of cells in the **pars distalis,** called **acidophils and basophils,** are closely associated with fenestrated capillaries.
   b. Acidophils
      (1) Acidophils, pinkish-staining cells, produce **growth hormone,** also called **somatotropin,** which acts on the epiphyseal plate and affects growth of all organs.
      (2) **Prolactin**, a product of these cells, stimulates lactation of mammary glands.
   c. Basophils
      (1) Basophils, purple-staining cells, produce **thyroid-stimulating hormone** (TSH), which affects the thyroid gland.

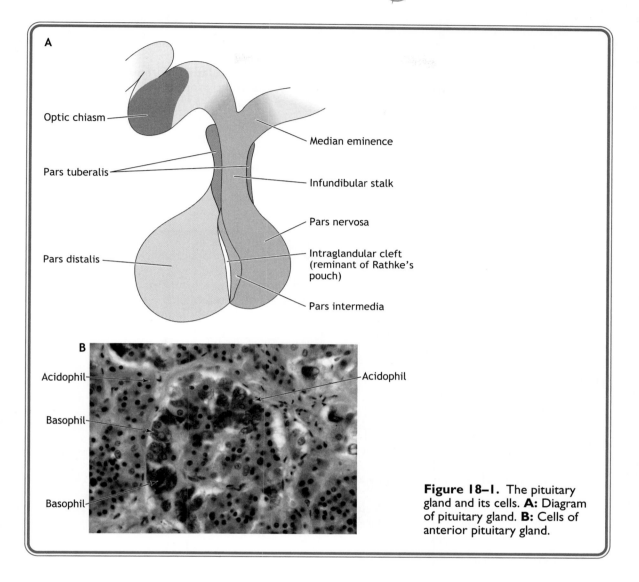

A

Optic chiasm

Pars tuberalis

Pars distalis

Median eminence

Infundibular stalk

Pars nervosa

Intraglandular cleft
(reminant of Rathke's
pouch)

Pars intermedia

B

Acidophil

Basophil

Basophil

Acidophil

Figure 18–1. The pituitary
gland and its cells. **A:** Diagram
of pituitary gland. **B:** Cells of
anterior pituitary gland.

(2) **Luteinizing hormone** (LH) stimulates formation of the corpus luteum,
progesterone secretion, ovulation, and secretion of testosterone by interstitial cells of testes.

(3) **Follicle-stimulating hormone (FSH)** stimulates growth of ovarian follicles and stimulates Sertoli cells to synthesize and secrete androgen-
binding protein.

(4) **Adrenocorticotropic hormone** (ACTH) affects the 3 zones of the
adrenal cortex and is formed from cleavage of the large precursor molecule called **pro-opiomelanocortin** (POMC).

2. Pars intermedia
      a. **Rathke's cysts** are colloid-containing cysts lined by a cuboidal epithelium found within the pars intermedia.
      b. Cells of the pars intermedia produce **melanocyte-stimulating hormone** (MSH) and **β-endorphin,** which form by enzymatic cleavage of POMC.

   D. **Neurohypophysis (Posterior Pituitary Gland)**
      1. General features
         a. The **neurohypophysis** consists of irregular lobules that contain neurosecretory axons from the hypothalamus and fenestrated capillaries.
         b. The neurohypophysis is formed of the **pars nervosa, infundibulum, and median eminence.**
      2. Pars nervosa (posterior lobe)
         a. **Pituicytes** are glial-like cells that contain aging pigment and account for 25% of the cells of the pars nervosa.
         b. Terminal axons, called **Herring bodies,** within the pars nervosa contain granules of 2 hormones and are a site of release and storage of hormones but not a site of their synthesis.
         c. **Antidiuretic hormone** (ADH), also called **vasopressin,** is synthesized in the supraoptic nucleus and paraventricular nuclei of hypothalamus. It functions to increase water resorption from collecting tubules in the kidney.
         d. **Oxytocin** is synthesized in the supraoptic and paraventricular nuclei of the hypothalamus and functions to cause contraction of uterine smooth muscle and myoepithelial cells in the mammary gland.

II. Thyroid Gland

   A. **General Features**
      1. The thyroid gland (Figure 18–2A) develops from the **foramen cecum** of the posterior segment of the tongue. The **thyroglossal duct** trails the thyroid gland as it migrates and degenerates. The duct can persist, connecting the cecum with the thyroid gland.
      2. The **thyroid gland** is covered by a capsule and consists of 2 lobes united at its isthmus.
      3. Septa from the capsule form follicle-containing **lobules.**
      4. The thyroid has an extensive blood and lymphatic supply, and fenestrated capillaries are found between follicles.

   B. **Thyroid Follicle**
      1. The functional unit of the thyroid gland is the **follicle,** which measures 50–500 μm in diameter and numbers 3 million in adults.
      2. The follicle is lined by a **simple cuboidal epithelium** that rests on a basal lamina.
      3. The lumen contains a homogeneous gelatinous substance called **colloid.**
      4. Less active follicles are large, full of colloid, and lined with a simple cuboidal epithelium. Active follicles are smaller with less colloid and lined with columnar epithelial cells.

   C. **Thyroid Epithelial Cells**
      1. Parafollicular cells
         a. **Parafollicular cells** are large and have a pale-staining cytoplasm and may be outside or part of the follicle.

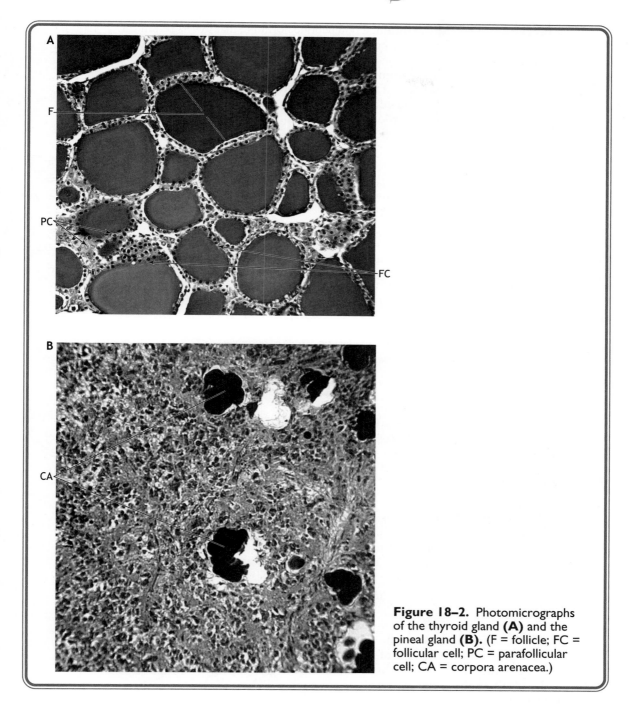

Figure 18–2. Photomicrographs of the thyroid gland **(A)** and the pineal gland **(B)**. (F = follicle; FC = follicular cell; PC = parafollicular cell; CA = corpora arenacea.)

b. These cells secrete **calcitonin,** a polypeptide hormone that lowers serum calcium levels by suppressing bone resorption by **osteoclasts.**

  2. Follicular cells

      a. **Follicular cells** are the main cells of the thyroid follicle.

      b. These cells, in response to TSH, synthesize **thyroglobulin,** which enters the colloid where it is **iodinated.**

      c. Follicular cells secrete **triiodothyronine** (T_3) and **thyroxin** (T_4), which function to increase basal metabolism.

HASHIMOTO'S DISEASE

- *Hashimoto's disease* results from an infiltration of the thyroid by lymphocytes, destroying the thyroid tissue and causing hypothyroidism.

GOITER

- *Goiter* is an enlargement of the thyroid gland, occurring with either hypothyroidism or hyperthyroidism.
- Goiter can be caused by ingesting goitrogenic foods, including cabbage, and intake of inadequate iodine, called *simple goiter.*

HYPOTHYROIDISM

- *Hypothyroidism* is marked by a decrease in the production of thyroid hormone.
- Clinical signs of hypothyroidism include unexplained weight gain, hard edema of connective tissue, and sleepiness.

HYPERTHYROIDISM

- *Hyperthyroidism* is seen in Grave's disease, in which a thyroid-stimulating antibody (TSAB) causes uncontrolled secretion of T_3 and T_4. The thyroid gland is no longer controlled by hormones released by the pituitary gland.

III. Parathyroid Gland

  A. General Features

    1. The parathyroid consists of **4 small glands** on the upper and lower poles of each thyroid lobe.

    2. These organs are separated from the thyroid by a capsule, and septa extend into the glands to support elongated, clusters of secretory cells.

  B. Secretory Cells

    1. Two main cells of the parathyroid are **chief** and **oxyphil cells;** adipose cells are also present.

    2. **Chief cells** are most numerous and the source of the **parathyroid hormone** (PTH), which is stimulated by a decrease in serum Ca^{2+}.

    3. PTH promotes **absorption of the calcified bone matrix,** thus elevating Ca^{2+} in the serum; aids in Ca^{2+} absorption from the gastrointestinal tract; and reduces PO_4 in the serum by increasing excretion of PO_4 in the kidney.

    4. **Oxyphil cells** are larger than chief cells and have an eosinophilic cytoplasm as a result of numerous mitochondria. These cells appear at about age 7 and increase in number with age.

PARATHYROID DISEASE

- **Hyperparathyroidism** *can be primary, secondary, or, rarely, tertiary. This disorder is marked by* **hypercalcemia.** *Primary hyperparathyroidism is a common endocrine disease caused by adenomas or hyperplasia.*
- **Hypoparathyroidism,** *less common than hyperparathyroidism, results in* **hypocalcemia.** *This disorder is affected by decreased secretion of PTH as a result of surgical removal of the parathyroid hormones during* **thyroidectomy.**

IV. Pineal Gland
A. General Features
1. The **pineal gland** (Figure 18–2B) is a conical gray body attached to the roof of diencephalon at the posterior extremity of the third ventricle.
2. A thin **capsule** of the pineal gland is continuous with the pia mater and arachnoid.
3. The pineal gland consists of cords of pale-staining **epithelial cells** arranged in lobules demarcated by connective tissue.
4. Cells of the pineal gland secrete **melatonin,** which affects circadian rhythms.

B. Cells
1. **Pinealocytes** have large nuclei, prominent nucleoli, and thin processes.
2. **Interstitial cells** are glial-like cells, with elongated nuclei, which are found in perivascular areas and between pinealocytes.
3. The pineal gland contains **corpora arenacea,** which are extracellular **concretions** of calcium phosphate and carbonate that increase with age.

V. Adrenal Gland
A. General Features
1. **Adrenal glands** (Figure 18–3A) are positioned at the cranial pole of each kidney.
2. Each gland has an outer **cortex** and an inner **medulla,** which are distinctive structurally, developmentally, and functionally.
3. A **capsule** surrounds each gland and sends septa into its cortex.

B. Blood Supply
1. Each adrenal gland is supplied by arteries that enter at various points along its periphery.
2. The arterial supply to the adrenal gland is supplied by the **superior suprarenal artery** from the inferior phrenic artery, the **middle suprarenal artery** from the aorta, and the **inferior suprarenal artery** from the renal artery.
3. These arteries form the subcapsular arterial plexus, which gives rise to capsular arteries.
   a. The arteries of the cortex that branch to form the cortical capillaries drain into the medullary capillaries.
   b. The arteries of the medulla that branch only when they reach the medulla form the **medullary capillaries.**
4. Medullary blood supply
   a. The **medullary capillaries** receive blood from arteries of the cortex and medulla that converge to form several medullary veins.

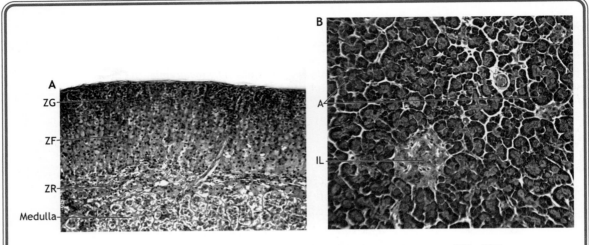

Figure 18–3. Photomicrographs of adrenal gland **(A)** and endocrine pancreas **(B).** (ZG = zona glomerulosa; ZF = zona fasciculata; ZR = zona reticularis; IL = islet of Langerhans; A = acinus.)

   b. **Medullary veins** ultimately converge to form the left suprarenal vein, which drains into the renal vein, and the right suprarenal vein, which drains into the inferior vena cava.
C. **Adrenal Cortex**
   1. Zona glomerulosa
      a. The **zona glomerulosa** is a thin, outer zone adjacent to the capsule. It makes up about 15% of the total cortical area.
      b. This region consists of columnar epithelial cells arranged in ovoid groups surrounded by capillaries.
      c. Cells in this region secrete **mineralocorticoids,** including **aldosterone,** which is involved in fluid and electrolyte balance.
   2. Zona fasciculata
      a. The **zona fasciculata** consists of large, polyhedral cells arranged in long cords radially arranged with respect to the medulla. It makes up about 65% of the total cortical area.
      b. This region has cords that are separated by sinusoidal blood vessels.
      c. Cells, called **spongiocytes,** in this region have extensive lipid droplets and smooth endoplasmic reticulum (SER) in cytoplasm.
      d. These cells secrete **glucocorticoids,** including **cortisol,** which regulate carbohydrate, protein, and fat metabolism.
   3. Zona reticularis
      a. The **zona reticularis** has an anastomosing network of cell cords that are continuous with the medulla. It makes up about 10% of the total cortical area.
      b. The lipofuscin-containing cells in this region accumulate with age.
      c. These cells secrete **glucocorticoids** and **sex hormones,** including **estradiol** and **testosterone.**

CUSHING'S SYNDROME

- *Cushing's syndrome* results from elevated secretion of cortisol and androgens.
- It causes diabetes mellitus, obesity, and high blood sugar levels.

CONN'S SYNDROME

- *Conn's syndrome* results from increased secretion of aldosterone, which causes retention of water and sodium and increased blood pressure.

ADDISON'S DISEASE

- *Addison's disease* is caused by a chronic hypofunctional adrenal cortex.
- This disease results in weakness, weight loss, low serum glucose, and high potassium.

    D. Adrenal Medulla
        1. The **adrenal medulla** is composed of large sympathetic ganglion cells, called chromaffin cells, arranged in rounded groups or short cords.
        2. Chromaffin cells, derived from the neuroectoderm, secrete **catecholamines,** such as **norepinephrine** and **epinephrine,** and are innervated by preganglionic sympathetic fibers.
        3. Epinephrine increases heart rate and cardiac output and elevates blood sugar levels by stimulating glycogen breakdown.
        4. Norepinephrine increases blood pressure via vasoconstriction and increases lipolysis.

TUMORS OF ADRENAL MEDULLA

- Tumors of **chromaffin cells** secrete elevated levels of catecholamines, which may cause a sustained stress response. Tumors of **ganglion cells** may also be present in the adrenal medulla, especially in children. The clinical signs of this disorder vary.

VI. Endocrine Pancreas

    A. General Features
        1. The **endocrine portion** of the pancreas consists of about 500,000 islets of Langerhans that are randomly distributed in the adult pancreas (Figure 18–3B).
        2. These **islets** make up 1–2% of the gland and are arranged in irregular cords of polyhedral endocrine cells.
        3. Islets consist of at least **6 endocrine cell types,** but alpha, beta, and delta are best characterized.
    B. Alpha Cells
        1. Alpha (A) cells comprise **15–20%** of the islet and are located peripherally in the islet.
        2. These cells produce **glucagon,** which functions to decrease uptake of glucose by cells.
        3. These cells contain dense secretory vesicles embedded in a less dense material.
    C. Beta Cells
        1. Beta (B) cells comprise **70%** of the islet and are centrally located in the islet.

2. These cells are **insulin**-producing cells and function to increase glucose uptake by cells.

3. These cells contain secretory vesicles that have a dense core of rhomboidal or polygonal crystals.

D. Delta Cells

1. Delta (D) cells comprise **5–10%** of this tissue and are located peripherally in the islet.

2. These cells secrete **somatostatin** or **gastrin.** Somatostatin suppresses insulin, glucagon, and growth hormone secretion by cells of the endocrine pancreas, whereas gastrin stimulates secretion of HCl by parietal cells of the gastric mucosa.

3. These cells contain secretory vesicles with a homogenous granular matrix.

DIABETES

· *Diabetes mellitus* is a disease caused by a reduction in the level or complete lack of insulin, which leads to reduction in metabolism of carbohydrates and increased utilization of lipids and protein. *Type I diabetes* is insulin dependent and accounts for 10% of all cases, whereas *Type II diabetes* is noninsulin dependent, accounting for 90% of cases.

· *Diabetes insipidus* is caused by decreased secretion of ADH by the anterior pituitary gland. This condition leads to dehydration and thirst as a result of the excretion of a large volume of urine.

CLINICAL PROBLEMS

A computed tomography (CT) scan revealed a tumor within the hypophysis of the pituitary gland of a 56-year-old man. It was determined that the man suffers from Cushing's disease.

1. Secretion of which of the following hormones would be elevated?

A. Adrenocorticotropic hormone (ACTH)

B. Follicle-stimulating hormone

C. Luteinizing hormone

D. Thyroid-stimulating hormone

E. Growth hormone

A patient suffers from a pituitary disorder, resulting in a reduction in adrenocorticotropic hormone (ACTH) production.

2. Which of the following would be reduced in secretion by the zona fasciculata of the adrenal gland?

A. Aldosterone

B. Renin

C. Mineralocorticoids

D. Epinephrine

    E. Cortisol

3. Which of the following occurs in response to glucagon secretion by the islets of Langerhans?

    A. Decreased blood glucose levels

    B. Decreased sodium ion uptake by proximal convoluted tubules

    C. Increased water retention by renal collecting tubules

    D. Elevated blood glucose levels

    E. Inhibition of hormone secretion by the adrenal cortex

4. Iodination of thyroglobulin within the thyroid gland occurs at which of the following sites?

    A. Parafollicular cells

    B. Follicle cells

    C. Colloid

    D. Basement of follicular cells

    E. Capillaries of thyroid gland

5. Which of the following hormones are secreted from beta (B) cells of the pancreas?

    A. Glucagon

    B. Insulin

    C. Somatostatin

    D. Secretin

    E. Vasoactive intestinal polypeptide

6. Which of the following can remain tethered to the back of the tongue after birth?

    A. Thyroid gland

    B. Adrenal gland

    C. Parathyroid gland

    D. Thymus

    E. Pancreas

A pathologist is studying a section of the pituitary gland. He notes a highly vascularized region of this gland. Many of the cells within this region have a basophilic cytoplasm.

7. Which of the following regions of the pituitary gland is he examining?

    A. Infundibular stalk

    B. Pars nervosa

    C. Pars intermedia

    D. Pars distalis

    E. Pars tuberalis

8. Based on question 7, which of the following hormones is released from the basophilic-staining cells?

   A. Antidiuretic hormone

   B. Prolactin

   C. Luteinizing hormone

   D. Oxytocin

   E. Somatomedin

9. Glial-appearing cells, called pituicytes, are found within which region of the pituitary gland?

   A. Pars tuberalis

   B. Pars nervosa

   C. Pars distalis

   D. Pars intermedia

   E. Infundibular stalk

10. The secretion of which of the following hormones is controlled by a feedback mechanism directed through the vasculature of the hypothalamus?

    A. Thyroid-stimulating hormone

    B. Insulin

    C. Calcitonin

    D. Oxytocin

    E. Antidiuretic hormone

ANSWERS

1. The answer is A. ACTH is synthesized and secreted from the hypophysis of the pituitary gland.

2. The answer is E. The zona fasciculata produces glucocorticoids, including cortisol.

3. The answer is D. Glucagon increases blood glucose levels.

4. The answer is C. Iodination of thyroglobulin is performed within the colloid of a thyroid follicle.

5. The answer is B. Beta cells of islets of Langerhans within the pancreas secrete insulin.

6. The answer is A. The thyroid gland develops from the foramen cecum at the back of the tongue. As the thyroid gland migrates, it can remain attached to the thyroglossal duct, which normally degenerates.

7. The answer is D. The pars distalis contains an abundant blood supply to facilitate releasing factors reaching the cells to signal secretion of hormones.

8. The answer is C. Basophils secrete 4 polypeptide hormones, including luteinizing hormone. Somatomedin and prolactin are secreted from acidophils, whereas oxytocin and ADH are released from cells of the pars nervosa.

9. The answer is B. The pars nervosa consists primarily of small glial-like cells called pituicytes.

10. The answer is A. Thyroxine secretion is controlled by thyrotropin, also called thyroid-stimulating hormone (TSH). TSH is produced within basophils of the pars distalis and is negatively affected by a feedback mechanism from the hypothalamus.

CHAPTER 19
FEMALE REPRODUCTIVE SYSTEM

I. Ovary
  A. General Features
  1. The ovary (Figure 19–1A) is covered by a simple squamous or cuboidal epithelium called the **germinal epithelium.**
  2. Beneath the epithelium is a layer of dense connective tissue called the **tunica albuginea.**
  3. The ovary has a thick **cortex** containing oocytes that surrounds its **medulla.**

  B. Ovarian Follicles
  1. **Primordial follicles** are unilaminar and quiescent and number about 500,000 at birth (Figure 19–1B).
     a. These follicles have a large nucleus with a prominent nucleolus and can degenerate to become **atretic follicles.**
     b. A single layer of flattened **follicular (granulosa) cells** surrounds a follicle.
  2. **Primary follicles** are stimulated by **follicle-stimulating hormone (FSH)** to transition from a primordial follicle.
     a. **Oocytes** enlarge, and the single layer of follicular cells becomes cuboidal to form a **unilaminar primary follicle.**
     b. Mitotic activity of cells in this layer results in a **multilaminar primary follicle.**
     c. The **zona pellucida** is deposited around more advanced **primary oocytes.** Both the oocyte and follicular cells contribute to its synthesis.
     d. The **theca folliculi** also develops, which includes a highly vascular **theca interna** and a **theca externa,** which is mainly connective tissue.
     e. The follicular cell layer remains **avascular** during growth of the follicle.
  3. Secondary follicle
     a. Follicle cells continue to proliferate, and the entire follicle enlarges to form **secondary follicles.**
     b. When the **granulosa layer** consists of 6–12 layers of follicle cells, irregular spaces filled with fluid begin to appear.
     c. These spaces increase in size and become confluent to form the **follicular antrum,** which is filled with **liquor folliculi.**
     d. The oocyte reaches its full size at this stage, but follicle cells continue to grow.
  4. Graafian follicle
     a. Approximately 10–14 days are required for the **primordial follicle** to reach a mature **graafian follicle.**

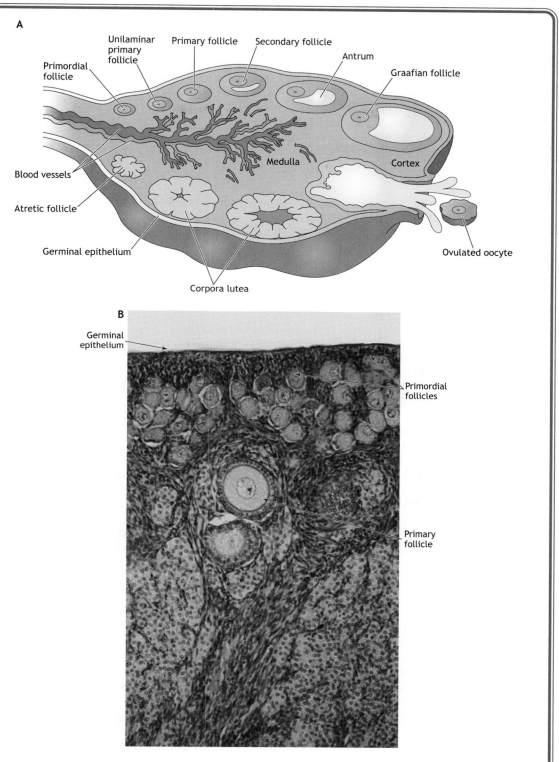

Figure 19–1. Diagram **(A)** and photomicrograph **(B)** of ovary.

b. This follicle bulges from the free surface of the ovary.

c. The oocyte is still attached to wall of the follicle by the **cumulus oophorus,** a small collection of granulosa cells.

d. At this stage, cells of the **theca interna** reach maximum development and resemble steroid-secreting cells.

e. Theca interna cells secrete **androstenedione,** which is converted to **17β-estradiol** by granulosa cells.

f. The **corona radiata** surrounds the **graafian follicle.**

C. Ovulation

1. **Ovulation** occurs in the middle of the **menstrual cycle** about every 28 days; however, variations of 7 days are common.

2. One oocyte is released while other developing follicles undergo degeneration.

3. Immediately before ovulation, the oocyte and attached granulosa cells float free in the follicle fluid.

D. Formation of Corpus Luteum

1. The **corpus luteum** is formed by **luteinizing hormone** (LH) stimulation.

2. It consists of remaining granulosa cells and cells of the theca interna, which are now called **granulosa lutein cells** and **theca lutein cells.**

3. The wall of the follicle collapses and is thrown into folds.

4. The corpus luteum secretes **progesterone,** which prevents the development of new ovarian follicles.

5. If an oocyte is **not fertilized,** the **corpus luteum of menstruation** is formed, which lasts about 14 days. It becomes a **corpus albicans** and gradually disappears.

6. If an oocyte is **fertilized** and **implants,** it becomes the **corpus luteum of pregnancy** and persists for about 6 months.

OVARIAN NEOPLASMS

· *Ovarian cancers* are common forms of neoplasms (6%) in females. The most common ovarian neoplasm (70%) is derived from the ovarian *surface epithelium.*

· Neoplasms of *germ cells* account for 15–20% of these ovarian tumors. *Sex cord-stromal tumors* account for 5–10% of ovarian neoplasms. These tumors develop from theca cells or components of the ovarian stroma.

II. Uterine Tube

A. General Features

1. The **uterine tube,** also called the oviduct (Figure 19–2A), is composed of an **infundibulum,** an expanded **ampulla,** and a terminal **isthmus.**

2. From the isthmus, the uterine tube enters the uterus at the superolateral region described as the **intramural region of the uterine tube.**

3. The end of the uterine tube has numerous finger-like processes called **fimbriae,** which function to receive the ovulated ovum.

4. This muscular tube provides communication between the uterine and peritoneal cavities.

5. The oviduct provides an environment for **fertilization** and initial stages of embryo development and transports this embryo to the uterus.

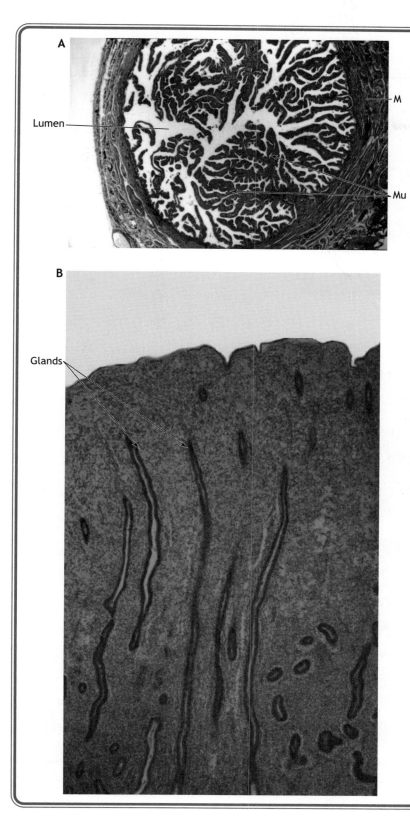

Figure 19–2. Photomicrographs of **(A)** uterine tube (M = muscularis; Mu = mucosa) and **(B)** uterus. (continued)

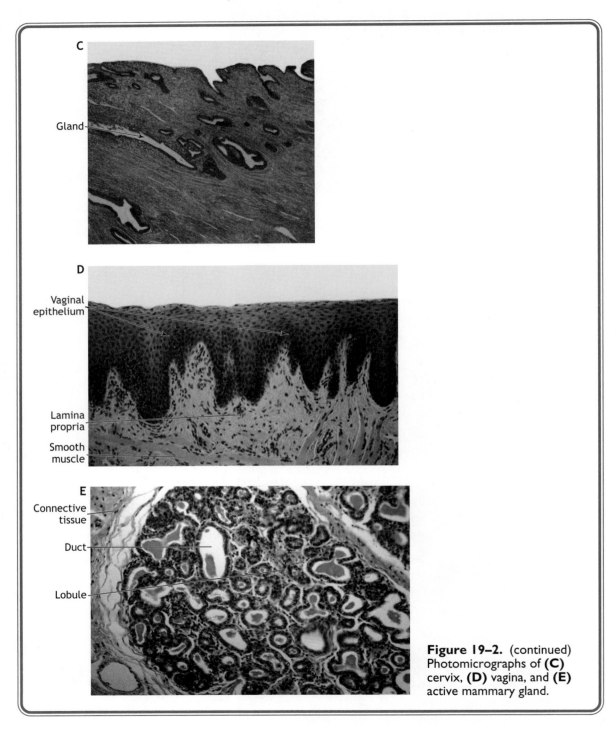

C

Gland

D

Vaginal
epithelium

Lamina
propria

Smooth
muscle

E

Connective
tissue

Duct

Lobule

Figure 19–2. (continued) Photomicrographs of **(C)** cervix, **(D)** vagina, and **(E)** active mammary gland.

B. **Histology**
1. The wall of the uterine tube consists of a mucosa, muscularis, and external serosa.
2. Mucosa
   a. The **mucosa** is highly folded in the **ampulla,** but folding decreases toward the uterus in the **isthmus.**
   b. Ciliated, simple columnar and nonciliated simple columnar cells, which are **secretory cells,** line the uterine tube.
   c. **Ciliated cells** are more numerous in the **ampulla** and less numerous in the **isthmus** and **intramural region.**
3. The muscularis consists of an inner circular layer and an outer longitudinal layer of smooth muscle.

UTERINE TUBE LESION

- *The most common lesion of the uterine tube is a **paratubal cyst** such as a **hydatid of Morgagni,** which is found at the initial segment of the tube.*

UTERINE MESOTHELIOMA

*Tumors of the uterine tube are rare, although **mesotheliomas** can occur beneath the serosa.*

III. Uterus

A. **General Features**
1. Functions of the uterus
   a. Cells of the uterus secrete substances that **nourish the embryo** before attachment and establish a favorable environment after attachment (Figure 19–2B).
   b. The uterus releases the fetus at parturition and forms part of the placenta.
2. Uterine layers
   a. The outer **perimetrium** is a serosa or adventitia.
   b. The middle **myometrium** contains smooth muscle and large blood vessels.
   c. The innermost **endometrium** consists of glandular, secretory components.
B. **Myometrium**
1. Four smooth muscle layers of the **myometrium** are found in this layer.
2. The **myometrium** goes through a period of growth during pregnancy and a reduction in size after pregnancy.
3. In the nonpregnant state, the myometrium is constantly undergoing shallow contractions without sensations, which may be exaggerated during sexual stimulation or during menstruation.
4. In the pregnant state, **progesterone** inhibits myometrial contractility and decreases at parturition. **Oxytocin** and **prostaglandins** stimulate contraction.

LEIOMYOMA

- *A **leiomyoma** is a benign tumor primarily found within the **myometrium** of the uterus. This tumor is the most common tumor in females and may cause excessive menstrual bleeding.*

C. **Endometrium**
1. The **endometrium** consists of a surface epithelial layer and underlying connective tissue, which is under the hormonal control of the ovary.

a. The surface epithelium invaginates to form numerous **uterine glands.**
   b. The connective tissue is called the endometrial stroma, which resembles mesenchyme with irregularly stellate cells with large, ovoid nuclei.
2. The surface epithelium is composed of simple columnar secretory cells with scattered ciliated cells.
   a. **Uterine glands** are simple tubular cells with bifurcations.
   b. Glandular secretion is called uterine milk.

ENDOMETRIOSIS

- *Endometriosis* is a condition in which endometrial elements are found within the pelvic cavity, such as the ovary and broad ligament of uterus.
- *Women suffering from endometriosis have painful menstruation.*

ENDOMETRIAL CARCINOMA

- *Endometrial carcinoma occurs most commonly in women older than 55 years.*
- *This form of carcinoma develops from excessive* **estrogen** *exposure and hyperplasia of the endometrium.*

   D. **Menstrual Cycle**
      1. General features
         a. The ovarian hormones **estrogen** and **progesterone** cause the endometrium to undergo cyclic, structural changes.
         b. These cycles begin at puberty and continue until menopause.
      2. Proliferative phase
         a. The **proliferative phase** is preceded by the menstrual phase and occurs between days 5–14.
         b. It takes place during ovarian follicular development.
         c. The basal layer of the **endometrium** remains after menstruation and proliferates. **Uterine gland cells** proliferate, migrate to the surface, and reconstitute the epithelium.
         d. The **glands** are straight and narrow at the end of this phase, whereas **coiled arteries** are elongated and convoluted.
      3. Secretory or luteal phase
         a. The **secretory phase** begins at ovulation and is dependent on **progesterone** produced by the corpus luteum.
         b. This phase occurs from day 15 to day 28 of the cycle.
         c. During this phase, the **endometrium** reaches its maximum thickness and **uterine glands** become tortuous, dilated, and secretory.
         d. Elongation and convolution of the coiled arteries continue and extend into the superficial portion of the endometrium.
         e. **Progesterone** stimulates the glands to secrete **glycoproteins,** which will be the major source of embryonic nutrition before implantation occurs.
      4. Menstrual phase
         a. The beginning of **menstrual blood** signals the initial stages of the **menstrual cycle.**
         b. This phase occurs when implantation fails and **estrogen** and **progesterone** levels fall.

c. At the end of the **secretory phase,** the walls of the coiled arteries contract, causing **ischemia** and necrosis of the **endometrium** followed by desquamation of the endothelium and rupture of blood vessels.

d. By the end of the menstrual phase, only the **basal layer** of the endometrium remains and the **proliferative phase** gradually restores the endometrium.

IV. Cervix

A. General Features
1. The cervix (Figure 19–2C) is a narrow extension of the uterus.
2. The cervix projects into the vagina.

B. Histology
1. The mucous membrane of the cervix is folded and consists of a simple columnar epithelium and lamina propria.
2. Numerous large **branched tubular glands** are present that consist of tall mucus-secreting columnar cells.
3. The external segment of the cervix that bulges into the lumen of the vagina is covered by stratified squamous, nonkeratinized epithelium.
4. The cervix **dilates** at parturition to accommodate the fetus.
5. Smooth muscle and elastic fibers are not a major component of the cervical wall.
6. The hormone **relaxin,** secreted by the corpus luteum of pregnancy, softens the cervix by increasing blood supply and tissue fluid content.

CERVICAL TUMOR

• Cervical cancer has been linked to the *human papillomavirus (HPV). Invasive squamous cell carcinomas* develop from squamous intraepithelial lesions. *Invasive adenocarcinomas* develop from glandular intraepithelial lesions.

V. Vagina

A. General Features
1. The vagina (Figure 19–2D) extends from the cervix to the vestibule.
2. The vagina consists of a mucosa, muscular layer, and adventitia.

B. Histology
1. **Mucosa** consists of a stratified squamous, nonkeratinized epithelium containing **glycogen** and a lamina propria.
2. The **acidity** within the vagina is due to the **fermenting** activity of bacteria on **glycogen** released into the lumen when epithelial cells are shed.
3. The **muscular layer** consists of smooth muscle oriented in interlacing bundles. Some are arranged circularly and others are longitudinal.
4. The **adventitia** consists of dense connective tissue containing an extensive venous plexus, sensory receptors, and nerve fibers.
5. The lamina propria and adventitia have abundant **elastic fibers.**

VAGINAL CANCER

• Vaginal carcinomas are rare tumors. These tumors are most frequently *squamous cell carcinomas.* This tumor usually arises on the superoposterior region of the vagina. *Adenocarcinomas* of the vagina are most frequently formed in the anterior region of the vagina. These 2 tumors are not easily treated.

VI. Mammary Gland

A. General Features
1. The mammary glands (Figure 19–2E) consist of modified **apocrine sweat glands,** which produce an exocrine secretion.
2. These glands are compound, tubuloalveolar glands.
   a. Each lobe is separated by dense connective tissue and contains **adipose tissue.**
   b. Each lobe extends a **lactiferous duct** that emerges at the nipple.

B. Inactive Mammary Gland
1. In an **inactive mammary gland,** the intralobular connective tissue is dense and abundant and contains adipose tissue.
2. The glandular ducts are lined by epithelial cells, but the few alveoli are small.

C. Lactating Mammary Gland
1. Soon after parturition, a **lactating mammary gland** secretes milk rich in fat, sugar, and proteins.
2. The alveoli become dilated with milk and have a low epithelium.
3. **Estrogen** and **progesterone** cause growth of the ducts at puberty.
4. During pregnancy, a continuous, prolonged secretion of these hormones, placental **lactogen,** and adrenal corticoids occurs.
5. **Oxytocin** secretion from the pars nervosa stimulates myoepithelial cell contraction and promotes **milk letdown.**
6. **Plasma cells** in the connective tissue surrounding the alveoli secrete immunoglobulin A into the milk, which provides the fetus with **passive immunity.**

MAMMARY GLAND TUMORS

- *Breast carcinoma* is a common, slow-growing, malignant tumor in women. Mutations of the tumor suppressor genes *BRCA1* and *BRCA2* (breast cancer genes) account for the majority of hereditary breast cancer (10% of all breast cancers). Mutation of the *BRCA1* gene indicates a high risk for ovarian, male breast, prostate, and pancreatic cancers.

CLINICAL PROBLEMS

1. Which of the following stages of follicular development is marked by an initial period of accumulation of follicular fluid?

   A. Graafian follicle

   B. Mature follicle

   C. Primordial follicle

   D. Oocyte

   E. Secondary follicle

2. Which of the following hormones is primarily responsible for inducing ovulation?

   A. Relaxin

   B. Luteinizing hormone

   C. Progesterone

   D. Follicle-stimulating hormone

   E. Estrogen

3. Which of the following produces progesterone?

   A. Granulosa lutein cells

   B. Cumulus oophorus

   C. Cells of corpus albicans

   D. Theca interna

   E. Theca externa

4. Which of the following characterizes the cumulus oophorus?

   A. Derived from cells of the theca interna

   B. Surrounds a secondary follicle

   C. Provides precursor cells to the corpus albicans

   D. Attaches the oocyte to wall of follicle

   E. Synthesizes and secretes progesterone

5. The ischemic phase of the menstrual cycle is triggered by a decrease in which of the following hormones?

   A. Prolactin

   B. Follicle-stimulating hormone

   C. Progesterone

   D. Luteinizing hormone

   E. Human chorionic gonadotropin

6. Which of the following is the first histological sign that a primordial follicle has matured to form a primary follicle?

   A. Appearance of spaces within granulosa cells

   B. Follicle cells change from squamous to cuboidal

   C. Presence of a theca interna and a theca externa

   D. Formation of multiple layers of follicle cells

   E. Appearance of a mature zona pellucida

7. The hormone 17β-estradiol is formed by conversion of androstenedione within which of the following?

   A. Granulosa cells

   B. Theca interna

   C. Oocyte

D. Antrum

E. Zona pellucida

8. Which of the following regions of the uterine tube have the greatest amount of folding of the mucosa?

A. Fimbria

B. Isthmus

C. Infundibulum

D. Ampulla

E. Intramural region

ANSWERS

1. The answer is E. The stage of oocyte development in which follicular fluid begins to accumulate is the secondary follicle.

2. The answer is B. Luteinizing hormone (LH) stimulates ovulation.

3. The answer is A. Granulosa interna cells of the corpus luteum secrete progesterone.

4. The answer is D. The cumulus oophorus is a small hillock of granulosa cells that attaches the mature oocyte to the wall of the follicle.

5. The answer is C. The ischemic phase of the menstrual cycle is triggered by a decrease in the blood level of progesterone. A decrease in estrogen within the blood also plays a role in the initiation of this phase.

6. The answer is B. The most noticeable change in the process of the maturation of a primordial follicle to the primary follicle is the follicle layer of cells becoming cuboidal from their initial squamous appearance. This stage is termed unilaminar follicle. Although the zona pellucida is beginning to form at the primary follicle stage, it is not mature until the multilaminar stage. The antrum, or spaces between the granulosa cells, does not occur until the secondary follicular stage.

7. The answer is A. Androstenedione is secreted by cells of the theca interna. This precursor hormone is converted to 17β-estradiol within granulosa cells of primary follicles.

8. The answer is D. The mucosa of the ampulla shows the greatest level of folding, which decreases as the uterine tube approaches the uterus.

CHAPTER 20
MALE REPRODUCTIVE SYSTEM

I. Testis

A. General Features

1. The **testis** (Figure 20–1) is suspended within the scrotum and surrounded by a testicular capsule, composed of 3 layers.
2. The **tunica vaginalis** is a layer of mesothelial cells consisting of a parietal and a visceral layer that surround the lateral, posterior, and anterior surfaces of the testis.
3. Tunica albuginea
   a. The **tunica albuginea** is a thick layer of dense connective tissue with scattered smooth muscle that immediately surrounds the testis.
   b. The tunica albuginea is thickened along the posterior surface of the testis, where it projects into the gland as the **mediastinum testis.**
4. The **tunica vasculosa** is immediately internal to the tunica albuginea.
5. Thin connective tissue partitions, called **testicular septa,** extend radially from the **mediastinum testis** to the capsule, dividing the interior of the testis into **250 pyramidal lobules.**

HYDROCELE

- **Hydrocele** is described as a collection of serous fluid within the tunica vaginalis, spermatic cord, or scrotum; the latter is called **Dupuytren's hydrocele.** Blood within the tunica vaginalis is called **hematocele.**

B. Seminiferous Tubules

1. General features
   a. Each lobule of the testis is composed of 1–4 highly convoluted **seminiferous tubules** (Figure 20–2A and B).
   b. Each seminiferous tubule is about 150 μm in diameter and 80 cm long.
   c. At the apex of a lobule, tubules lose their convolutions and become **tubuli recti.**
2. Seminiferous epithelium
   a. Tubules of the testis are lined by a **seminiferous epithelium** that has a stratified appearance.
   b. The 2 major cell types are supportive **Sertoli cells** and **spermatogenic cells.**
3. Sertoli cells
   a. Columnar **Sertoli cells** rest on a basal lamina and extend the full thickness of the epithelium.

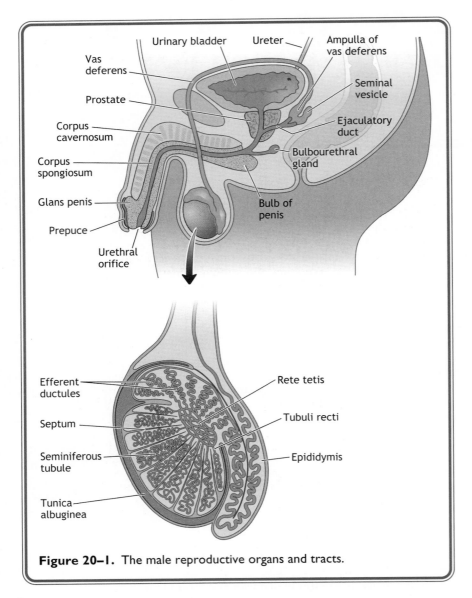

Figure 20–1. The male reproductive organs and tracts.

b. These cells are support cells for spermatogenic cells. They enable movement of spermatogenic cells from the basal lamina to the lumen and provide nutrition for sperm.

c. **Sertoli cells** have an oval nucleus with a nucleolus, a smooth endoplasmic reticulum (SER), numerous lysosomes, and Golgi apparatus.

d. Sertoli cells are joined by tight junctions to form the **blood–testis barrier.**

e. These cells release late **spermatids** into the lumen and **phagocytize** residual cytoplasm after release of late spermatids as well as aberrant germ cells.

f. Sertoli cells secrete **androgen-binding protein** (ABP) and **inhibin.**

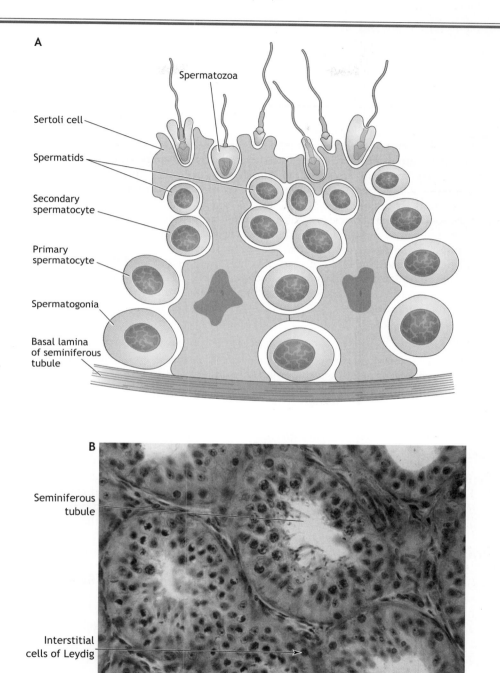

Figure 20–2. Diagram **(A)** and photomicrograph **(B)** of the seminiferous tubules.

4. Germ cells
   a. The 5 types of germ cells are formed during the process called **spermatogenesis.**
   b. **Spermatogonia** are about 12 μm in diameter and border the basal lamina of the seminiferous tubule. These cells have 46 chromosomes and 4N DNA and thus are diploid cells.
   c. Nuclei of these cells are round to ovoid in shape, with prominent nucleoli and a lightly stained cytoplasm.
   d. The **A spermatogonia** serve as stem cells and undergo mitosis slowly, whereas **B spermatogonia** are committed to the production of **spermatocytes.**
   e. **Primary spermatocytes** are the largest germ cells and concentrate in the middle of the epithelium. These cells have 46 chromosomes and 4N DNA and thus are diploid cells.
   f. **Secondary spermatocytes** result from the first meiotic division and have 23 chromosomes and 2N DNA. These cells have about one half the volume of primary spermatocytes and lie nearer to the lumen.
   g. **Spermatids** are haploid cells with 23 chromosomes and 1N DNA and arise from division of secondary spermatocytes. These cells lie close to the lumen and do not undergo further division but are transformed into **spermatozoa** during **spermiogenesis.**

CRYPTORCHIDISM

- *Undescended testes result from a failure of the abdominal testes to descend into the scrotum. This condition is called **cryptorchidism.** This anomaly is most often unilateral.*

SEMINOMA

- *The most common form of germ cell tumor in the male is the **seminoma** (50%), which is a malignant neoplasm usually affecting young males.*

C. **Peritubular Tissue**
   1. The tissue that surrounds the seminiferous tubules contains abundant connective tissue, flattened fibroblasts, and smooth muscle cells called **myoid cells.**
      a. The smooth muscle aids in the movement of spermatozoa along the length of the tubule.
      b. This tissue increases with age and may be extensive in certain cases of infertility.
   2. Blood vessels and nerves enter and leave from the **mediastinum** and form networks around the tubules.
   3. Interstitial cells of Leydig
      a. **Interstitial cells of Leydig** are found in groups, usually in the angular areas created by the seminiferous tubules.
      b. These cells are large vacuolated cells, with a distinct nucleus and nucleolus and SER. They serve as a source of **testosterone.**

D. **Hormonal Control**
   1. **Luteinizing hormone** (LH) controls the structure and endocrine function of **interstitial cells of Leydig.**

2. Testosterone is required to maintain spermatogenesis. It exerts a positive control on **Sertoli cells** and **germ cells.**

3. Testosterone also acts as an inhibitor of **LH** production and secretion by acting on the pituitary gland and hypothalamus.

II. Genital Ducts

A. Tubuli Recti

1. Seminiferous tubules at the apex of each lobule join to form **tubuli recti.**
2. Each tubule is **straight,** short, and devoid of convolutions.
3. At the junction of the seminiferous tubule, the spermatogenic cells disappear and only Sertoli cells remain.

B. Rete Testis

1. Tubuli recti course to the dense connective tissue of the mediastinum, where they enter a network of anastomosing channels called **rete testis.**
2. The rete testis appears as **irregular spaces** lined by simple cuboidal to columnar epithelium.
3. Passage of spermatozoa through the tubuli recti and the rete testis is rapid.

C. Efferent Ductules

1. In the superior portion of the posterior border of the testis, 10–15 spiral-oriented **efferent ductules** arise from the rete testis and emerge on the surface of the testis (Figure 20–3A).
2. These ductules are bound by connective tissue and surrounded by a thin layer of circularly arranged smooth muscle fibers.
3. Each ductule is 6–8 cm long and 0.05 mm in diameter.
4. These ductules are lined by a **simple columnar epithelium** that rests on a thin basal lamina.
   a. The lumen of the ductule is **irregular** because of the varying height of epithelial cells.
   b. Many of the taller cells have **motile cilia** that aid transport of spermatozoa to the epididymis.
   c. Nonciliated shorter cells have an **absorptive function.**

D. Epididymis

1. The highly, tortuous **epididymis** is surrounded by connective tissue and smooth muscle (Figure 20–3B).
2. The lumen is lined by a **pseudostratified epithelium** consisting of tall principal cells and smaller basal cells.
3. The **principal cells** have apical nonmotile **stereocilia** and a resorptive function.
4. **Basal cells** are small pyramidal cells that are **stem cells.** They replace principal and other stem cells.
5. Most of the fluid leaving the testis is reabsorbed in the **efferent ductules** and **epididymis.**
6. The epididymis serves as a site of **maturation** and **storage of sperm** and **phagocytosis** of defective sperm.

E. Ductus (Vas) Deferens

1. The epididymis straightens at its termination and becomes continuous with the **ductus deferens** (Figure 20–3C).

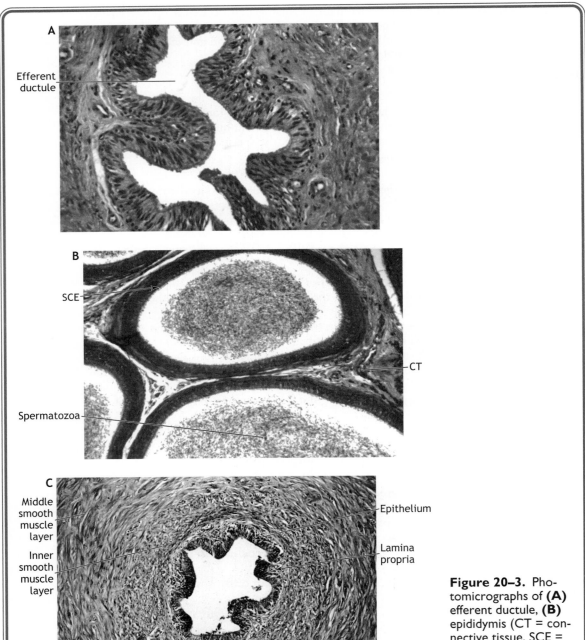

Figure 20–3. Photomicrographs of **(A)** efferent ductule, **(B)** epididymis (CT = connective tissue, SCE = simple columnar epithelium), and **(C)** vas deferens. (continued)

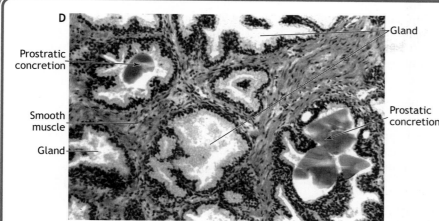

D

Gland

Prostratic
concretion

Smooth
muscle

Gland

Prostatic
concretion

Figure 20–3.
(continued)
Photomicrograph
of **(D)** prostate.

2. The ductus deferens ascends from the scrotum within the **spermatic cord** to the urethra. Before its termination, the duct dilates and enlarges into the **ampulla.**

3. The ductus deferens is lined by a **pseudostratified columnar epithelium,** and many of these cells have **stereocilia.**

4. The epithelium is supported by a lamina propria and a fibroelastic connective tissue.

5. The **muscular coat** consists of 3 distinct layers of smooth muscle that is surrounded by an external layer of fibroelastic connective tissue.

F. Ampulla
1. The **ampulla** receives the duct of the seminal vesicle. Together they form the **ejaculatory duct,** which passes through the substance of the prostate gland and drains into the **prostatic urethra.**

2. The epithelium of the ampulla is thickened and highly folded.

G. Spermatic Cord
1. The **spermatic cord** is enclosed by the **cremaster muscle.**

2. This cord contains the ductus deferens, spermatic artery, pampiniform plexus, and nerves.

3. A countercurrent exchange within vessels of the spermatic cord cools blood going to the testes.

III. Auxiliary Genital Glands

A. Seminal Vesicles
1. A **seminal vesicle** is a highly coiled tubular structure that joins the ampulla of the ductus deferens superior to the prostate gland.

2. The epithelium is **pseudostratified columnar epithelium,** consisting of columnar cells with microvilli and basal cells.

3. The seminal vesicle is an **exocrine gland** whose secretions represent 70% of semen. These secretions include prostaglandins, proteins, amino acids, and **fructose**; the latter is the energy source for **spermatozoa.**

B. Prostate

1. The **prostate** (Figure 20–3D) surrounds the urethra at its origin immediately inferior to the **urinary bladder.**
2. It consists of 30–50 small compound **tubuloalveolar glands** that drain into the **prostatic urethra** via 15–30 small excretory ducts.
3. The glands are found in the **mucosa, submucosa,** and **periphery** of the prostate.
4. The most peripheral **main glands** constitute the largest and most numerous of these glands.
5. Glands are lined by pseudostratified columnar epithelium with abundant rough endoplasmic reticulum (RER) and Golgi apparatus.
6. The lumen of these glands may contain prostatic concretions, called **corpora amylacea.**
7. Prostatic secretions
   a. The secretion of the prostate gland is colorless and slightly acidic and is regulated by **dihydrotestosterone.**
   b. Components of the secretion include **proteolytic enzymes,** amylase, citric acid, and **acid phosphatase.**

BENIGN NODULAR HYPERPLASIA

- In **benign nodular hyperplasia,** the stroma and glandular components of the prostate undergo hypertrophy. The cell increase leads to constricting flow of urine through the prostatic urethra.

CARCINOMA OF THE PROSTATE

- **Carcinoma** of the prostate is the most common cancer in males. The carcinoma usually arises in the posterior region of the prostate. Androgens may play a role in prostatic carcinoma.
- **Prostate-specific antigen** (PSA) and **prostatic acid phosphatase** are 2 serum markers of prostate cancer.

C. Bulbourethral Glands

1. **Bulbourethral glands,** also called Cowper's gland, lie within the sphincter urethrae muscle posterolateral to the membranous urethra.
2. The duct of these **compound tubuloalveolar glands** drains into the posterior segment of the bulbous portion of the **spongy urethra.**
3. Epithelium of these glands is simple columnar and secretes **galactose** and **sialic acid,** which has a role in lubrication.

IV. Penis

A. General Features

1. The **penis** (Figure 20–4) consists of paired **corpora cavernosa** and a **corpus spongiosum,** each surrounded by a fibrous capsule called the **tunica albuginea.**
2. More external, **Buck's fascia** covers the 3 cylinders and is, in turn, covered by a thin skin, having an abundant subcutaneous layer with smooth muscle.

B. Erectile Tissue

1. The erectile tissue of the **corpora cavernosa** is a spongelike system of irregular **vascular spaces** lined by endothelium, fed by efferent arteries, and drained by efferent veins.

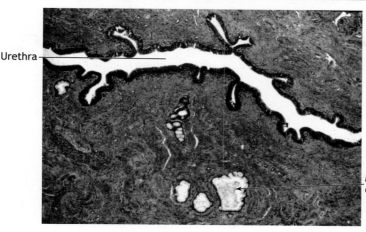

Urethra

Mucous glands
of Littre

Figure 20–4.
Photomicrograph
of penis.

2. Trabeculae separate vascular spaces and contain connective tissue and smooth muscle.
3. In the **flaccid state,** the vascular spaces contain little blood, which passes into the **arteriovenous anastomosis.**
4. In the **erectile state,** the vascular spaces become engorged with blood from the **dorsal and deep arteries of the penis.**
5. The shift of blood into the vascular spaces is controlled by the parasympathetic nervous system.
6. Blood drains from veins of the corpora cavernosa, corpus spongiosum, and glans penis into the **deep dorsal vein.**

C. Male Urethra
1. The **prostatic urethra** is surrounded by the prostate, lined by transitional epithelium, and receives ejaculatory ducts.
2. The **membranous urethra** is short, extends from prostate to the bulb of the corpus spongiosum, and is lined by stratified or pseudostratified columnar epithelium.
3. The **spongy urethra** (bulbous and penile portions), which courses through the corpus spongiosum, is lined by a stratified or pseudostratified columnar epithelium except at the terminal navicular fossa, where it is lined by stratified squamous epithelium.
4. **Glands of Littre** drain into the penile urethra.

CLINICAL PROBLEMS

While examining a cell line developed from the testis, you note mitotic cells at metaphase that have 46 chromosomes.

1. Which of the following cells are you examining?

   A. Mature sperm

   B. Secondary spermatocytes

   C. Spermatogonia

   D. Spermatids

2. Based on Question 1, in which of the following regions of the male genital ducts would you expect to find these cells?

   A. Within the epididymis

   B. At the basal lamina of the seminiferous tubule

   C. In the middle region of the seminiferous tubule

   D. At the adluminal region of the seminiferous tubule

   E. Within the ductus deferens

3. Which of the following serves as the site of final sperm maturation?

   A. Epididymis

   B. Seminiferous tubules

   C. Ampulla of ductus deferens

   D. Prostatic urethra

   E. Infundibulum of ductus deferens

4. As sperm pass through the male genital ducts, nutrients are provided by several sources. Which of the following provides a fructose-rich secretion?

   A. Interstitial cells of Leydig

   B. Cowper's gland

   C. Prostate gland

   D. Glands of Littre

   E. Seminal vesicles

5. Interstitial cells of Leydig play an important function in male gamete production. Because of this function, which of the following organelles is abundant within these cells?

   A. Lysosomes

   B. Smooth endoplasmic reticulum

   C. Peroxisomes

   D. Polyribosomes

   E. Golgi apparatus

6. Within the male reproductive tract, stereocilia project from cells lining which of the following regions?

    A. Rete testis

    B. Seminiferous tubules

    C. Ampulla of ductus deferens

    D. Epididymis

    E. Penile urethra

7. Which of the following organs within the male accumulate corpora amylacea with increasing age?

    A. Prostate

    B. Seminal vesicles

    C. Cowper's gland

    D. Epididymis

    E. Ductus deferens

8. Which of the following structures connect the rete testis to the epididymis?

    A. Ejaculatory duct

    B. Ductus deferens

    C. Efferent ductules

    D. Tubuli recti

    E. Spermatic cord

9. The small mucus-secreting glands of Littre secrete their product at which of the following sites?

    A. Prostatic urethra

    B. Corpora cavernosa

    C. Seminal vesicles

    D. Membranous urethra

    E. Penile urethra

10. Which of the following components of the male genital duct has a lumen that is irregular in appearance because of the varying heights of its epithelial lining?

    A. Rete testis

    B. Efferent ductules

    C. Ampulla of the ductus deferens

    D. Prostatic urethra

    E. Tubuli recti

ANSWERS

1. The answer is C. The only germ cell within the seminiferous tubules that has a diploid number of chromosomes (ie, 46) is the spermatogonia. All of the other germ cells are haploid, having 23 chromosomes.

2. The answer is B. The most immature cells, having 46 chromosomes, are found along the basal lamina, which supports the seminiferous tubules. The most mature sperm are found closer to the lumen of these tubules.

3. The answer is A. Sperm undergo final maturation or spermiogenesis within the epididymis.

4. The answer is E. The segment of the male reproductive tract that provides a fructose-rich secretion is the seminal vesicle.

5. The answer is B. The interstitial cells of Leydig synthesize and secrete the male steroid testosterone. The organelle that is responsible for the synthesis of steroids is the smooth endoplasmic reticulum.

6. The answer is D. The male duct lined with epithelium that projects stereocilia is the epididymis.

7. The answer is A. With aging, the prostate accumulates concretions called corpora amylacea within its parenchyma. These concretions consist of glycoproteins and calcium.

8. The answer is C. The efferent ductules connect the rete testis to the epididymis (at its head).

9. The answer is E. The ducts of glands of Littre drain into the penile urethra. These glands secrete a mucoid material.

10. The answer is B. The lumen of efferent ductules has an irregular appearance because the epithelial cells have varying heights. Some of the cells are ciliated, whereas others have absorptive functions and thus lack cilia.

CHAPTER 21
EYE AND EAR

I. Eye

A. General Features

1. The external layer, or **fibrous tunic,** of the eye consists of the sclera and cornea (Figure 21–1A).
2. The middle layer, or **vascular tunic,** consists of the iris, ciliary body, and choroid.
3. The internal layer, or **photoreceptive tunic,** is composed of the sensory retina.
4. The **refractile media** of the eye, from internal to external, consists of the cornea; anterior and posterior chambers, which contain aqueous humor; the lens; and vitreous space, which contains **vitreous humor.**

B. Fibrous Tunic

1. Cornea
   a. The **cornea,** the anterior one sixth of the eye, is avascular, colorless, and transparent and lacks lymphatics.
   b. The cornea consists of 5 layers.
      (1) The external **corneal epithelium** consists of a nonkeratinized stratified, squamous epithelium, usually 5–6 cell layers thick, that is continuous with the bulbar conjunctival epithelium.
      (2) **Bowman's membrane,** which is 7–12 μm thick, consists of fine collagen fibrils crossing at random and no cells.
      (3) The **stroma,** comprising the bulk (90%) of the cornea, is formed by layers of collagen bundles arranged at right angles to each other, an abundant ground substance rich in chondroitin sulfate, and a few cells.
      (4) **Descemet's membrane** is a homogeneous structure composed of fine collagen fibrils, which are 5–10 μm thick.
      (5) The innermost **corneal endothelium** is a simple squamous epithelium.
   c. Both the corneal epithelium and endothelium are responsible for maintaining the **transparency** of the cornea through their capability of transporting Na^+ to the apical surface, which keeps the stroma in a dehydrated state.
   d. The unique fiber arrangement of the cornea also contributes to its transparency.

CORNEAL DISORDER AND THERAPY

- **Cornea guttata,** also called dystrophia endothelialis corneae, is a degenerative condition of the cornea resulting from dystrophy of the endothelial cells.
- **Radial keratotomy** is an operation performed by making laser incisions of the cornea from the periphery to the center in a spoke-wheel fashion. This process flattens the cornea to correct for **myopia.**

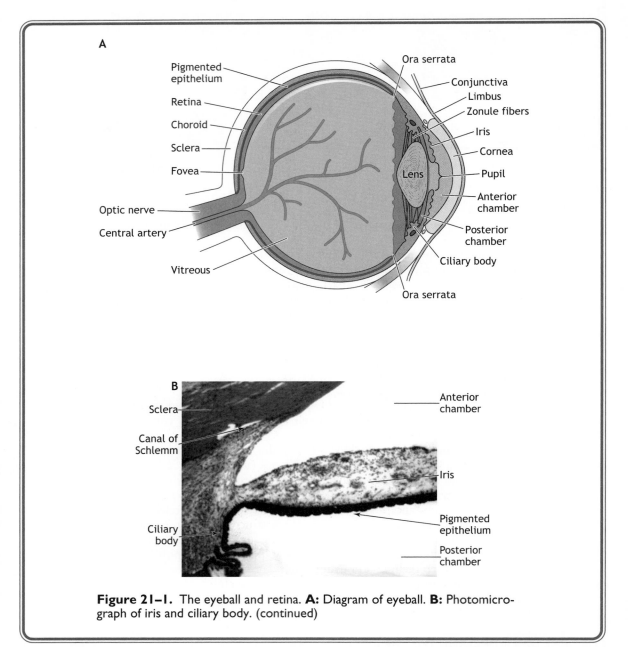

Figure 21–1. The eyeball and retina. **A:** Diagram of eyeball. **B:** Photomicrograph of iris and ciliary body. (continued)

2. Sclera
   a. The **sclera** is the posterior five sixths of the external layer of the eye, which maintains the size and shape of the eye.
   b. Its external surface appears grayish-white and is the site of attachment for recti and oblique ocular muscles.

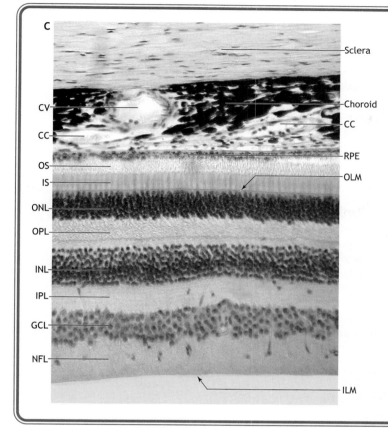

C

Sclera
Choroid
CV
CC
CC
OS
IS
ONL
OPL
INL
IPL
GCL
NFL
RPE
OLM
ILM

Figure 21–1. (continued) **C:** Photomicrograph of retina. (CV = choroidal vessel; CC = choriocapillaris; RPE = retinal pigment epithelium; OS = outer segment; IS = inner segment; OLM = outer limiting membrane; ONL = outer nuclear layer; OPL = outer plexiform layer; INL = inner nuclear layer; IPL = inner plexiform layer; GCL = ganglion cell layer; NFL = nuclear fiber layer; ILM = inner limiting membrane.)

  c. The **optic nerve** exits posteriorly at the **lamina cribrosa,** where the sclera is reduced to a fenestrated membrane.
  d. Anteriorly the sclera is covered by the **bulbar conjunctiva** and posteriorly by the bulbar sheath, called **Tenon's capsule.**
  3. Limbus
  a. The **limbus** is the transition between the cornea and sclera, which is highly vascularized with numerous nerve fibers.
  b. Within the stroma layer at the limbus, irregular, endothelium-lined channels, the **trabecular meshwork,** merge to form the **canal of Schlemm,** which drains aqueous humor from the anterior chamber of the eye.
C. Uvea
  1. Choroid
  a. The **choroid** is the vascularized coat, with loose connective tissue rich in fibroblasts, macrophages, lymphocytes, mast cells, plasma cells, and collagen and elastic fibers.
  b. The innermost layer, called the **choriocapillaris,** is rich in small blood vessels and provides nutrients to the retina.
  c. The choriocapillaris is separated from the retina by a thin membrane called **Bruch's membrane.**

2. Ciliary body and processes
   a. The **ciliary body** is an anterior expansion of the choroid that lies at the inner surface of the anterior portion of the sclera.
   b. It contains the **ciliary muscle,** which controls the curvature of the lens by exerting tension on zonule fibers.
   c. **Ciliary processes** are ridgelike extensions of the ciliary body from which the **zonule fibers** emerge.
   d. Cells from the ciliary processes actively transport certain constituents of plasma into the posterior chamber as **aqueous humor.**

DISORDER OF AQUEOUS HUMOR

- *Aqueous humor* within the anterior and posterior chambers of the eye exerts a pressure of 10–20 mm Hg. Blockage of aqueous humor flow at the **trabecular meshwork** in the anterior chamber results in an increased intraocular pressure, a condition called **glaucoma.** An increased and persistent pressure in excess of 25 mm Hg can cause damage to the optic disk and the ganglion cell axons that pass into the optic nerve, resulting in apoptotic death of **ganglion cells.**
- The most common form of glaucoma, **closed-angle glaucoma,** results from contact of the iris and the inner surface of the trabecular meshwork. This structural association prevents egress of the aqueous humor from the anterior chamber.
- In **opened-angle glaucoma,** the angle of the anterior chamber remains open; however, passage of aqueous humor through the **trabecular meshwork** gradually diminishes.

3. Iris
   a. The **iris** extends from the angle of the anterior chamber partially covering the lens, leaving a round opening in the center called the **pupil.** It separates the anterior and posterior chambers of the eye (Figure 21–1B).
   b. The anterior iris surface consists of a spongy stroma and contains fibroblasts, collagen fibers, and varying amounts of melanocytes.
   c. **Melanocytes** of the iris function to keep stray light from interfering with image formation and, in the stroma of the iris, are responsible for the color of the eye.
   d. The iris consists of an outer (anterior) layer composed of fibroblasts and melanocytes and an inner (posterior) double layer of pigmented epithelium. A stroma consisting of fibroblasts and connective tissue is found between these 2 layers.
   e. Smooth muscle fibers at the papillary margin form the **sphincter pupillae muscle,** which causes constriction under parasympathetic control.
   f. Basal processes of pigmented epithelial cells nearest to the stroma form the **dilator pupillae muscle,** which causes relaxation under sympathetic control.

UVEITIS

- *Uveitis* is a group of inflammatory disorders affecting a segment of the uvea, called **choroiditis, cyclitis,** or **iritis.** The agents causing the inflammatory response are certain viruses, fungi, parasites, and bacteria.

D. Photoreceptive Tunic
   1. The **retina** comprises the inner layer of the eyeball and consists of a posterior portion that is photosensitive (Figure 21–1C).

2. The **ora serrata** is the point of demarcation of the anterior and posterior portions of the retina.
3. The retina consists of 10 layers.
   a. **Retinal pigment epithelium** (RPE) is a simple cuboidal epithelium with a basal nucleus that serves a variety of supportive functions in the retina.
   b. The **photoreceptor layer** consists of outer and inner segments of rod and cones.
   c. The **outer limiting membrane** is a region of junctional complexes between Müller cell processes and photoreceptor cells.
   d. The **outer nuclear layer** contains cell bodies of rod and cone photoreceptor cells. Cones predominate at the macula lutea, whereas rods are most abundant in the peripheral retina.
   e. The **outer plexiform layer** consists of synapses between photoreceptors and bipolar cells.
   f. The **inner nuclear layer** contains cell bodies of Müller, bipolar, horizontal, and amacrine cells.
   g. The **inner plexiform layer** is the synaptic region between bipolar and ganglion cells.
   h. The **ganglion cell layer** contains retinal ganglion cells.
   i. The **nerve fiber layer** consists of unmyelinated nerve fibers of the ganglion cells that form the optic nerve.
   j. The **inner limiting membrane** is the basal lamina of Müller cell processes that separates the retina from the vitreous humor.
4. The **vitreous** is a hydrated gel that provides support for the posterior aspect of the lens. A few cells, particularly **hyalocytes** and **macrophages** and **type II collagen fibrils,** are found within the vitreous. The vitreous is firmly attached at the optic disk and Müller cell processes and maintains the attachment of the retina to the RPE.

RETINAL PIGMENT EPITHELIUM DISORDER

- *The **RPE** phagocytizes shed outer segments of photoreceptor cells and synthesizes and secretes several growth factors. With age, the RPE becomes defective in this supportive function, thus leading to death of photoreceptor cells, particularly cones, in the area of the fovea, termed **age-related macular degeneration** (ARMD). Loss of central vision is a symptom of ARMD.*

PHOTORECEPTOR CELL DISORDERS

- *The major protein of rod photoreceptor cells is **opsin,** which is a photopigment found primarily in outer segments. Point mutations of the **opsin** and **peripherin** gene result in rod photoreceptor cell death in **retinitis pigmentosa (RP).** This disease is inherited as an autosomal dominant recessive, maternally inherited trait or sex-linked recessive trait. **Night blindness** and loss of peripheral vision are symptoms of RP resulting from the loss of **rods.***

- *Separation of the retinal pigment epithelium from the sensory retina, called **retinal detachment,** can result in photoreceptor cell loss, leading to blindness. The separation of RPE from the retina can occur as a result of invading choroidal tumors (**exudative** retinal detachment), ocular trauma (**traction** detachment), or vitreous accumulation after a retinal tear (**rhegmatogenous** retinal detachment).*

    E. Lens
   1. The **lens** is the oval clear structure that functions to focus the light image at the fovea centralis.

2. The **lens capsule** consists of a basement membrane, 10–20 μm thick, consisting mainly of **type IV collagen** and glycoproteins.
3. The **subcapsular epithelium** is a cuboidal epithelium on the anterior surface of the lens.
4. **Lens fiber cells** are elongated, highly differentiated cells derived from epithelial cells. Eventually these cells lose their nuclei and other organelles. At maturation, lens fiber cells are 7–10 mm in length and contain **crystalline** proteins.

LENS DISEASE

• *Cataracts are diseases of the lens characterized by opacity of this structure, which impairs vision. The different forms of cataracts are classified by size, shape, location, and cause of opacity. The most common form of cataract is **senile cataract**, which develops with aging. Other causes of cataract formation include corticosteroid treatment, skin disorders, glaucoma, uveitis, and diabetes mellitus.*

II. Ear

A. External Ear
1. The **external acoustic meatus** is the canal that connects the outside environment to the tympanic membrane (Figure 21–2A).
2. This canal is lined by skin containing **ceruminous glands,** which secrete cerumen or earwax.

B. Middle Ear
1. The **middle ear** is found within the petrous portion of the **temporal bone** and extends from the tympanic membrane laterally to the oval and round windows medially.
2. The **malleus,** which attaches to the **tympanic membrane,** articulates with the **incus.** The **incus** articulates with the **stapes,** which inserts at the **oval window.**
3. The **stapedius muscle** attaches to the stapes and contracts in response to loud noises. The **tensor tympani muscle,** which is attached to the malleus, has a similar function.

MIDDLE EAR DISORDER

• *Bacterial infection of the middle ear, or **otitis media**, is a complication of infections of the respiratory tract leading to a severe earache. **Inflammatory cells** accumulate within the middle ear, applying pressure to or tearing the **tympanic membrane**. The infection can invade the mastoid air cells within the temporal bone, causing **mastoiditis**.*

C. Inner Ear
1. The bony cavity medial to the middle ear comprises that position of the inner ear called the **vestibule.**
2. An extension of the vestibule called the **cochlea** contains the membranous **cochlear duct** (or scala media), which contains the specialized epithelial organ of hearing, the **organ of Corti.**

D. Cochlea
1. Sound waves transmitted to the oval window by the **stapes** travel to the **scala vestibuli,** up through the turns of the cochlea to the apex, where it communicates with the **scala tympani** through the **helicotrema** (Figure 21–2B). These waves continue through to the end at the **round window. Perilymph,** containing high sodium, fills the scala tympani and vestibuli.

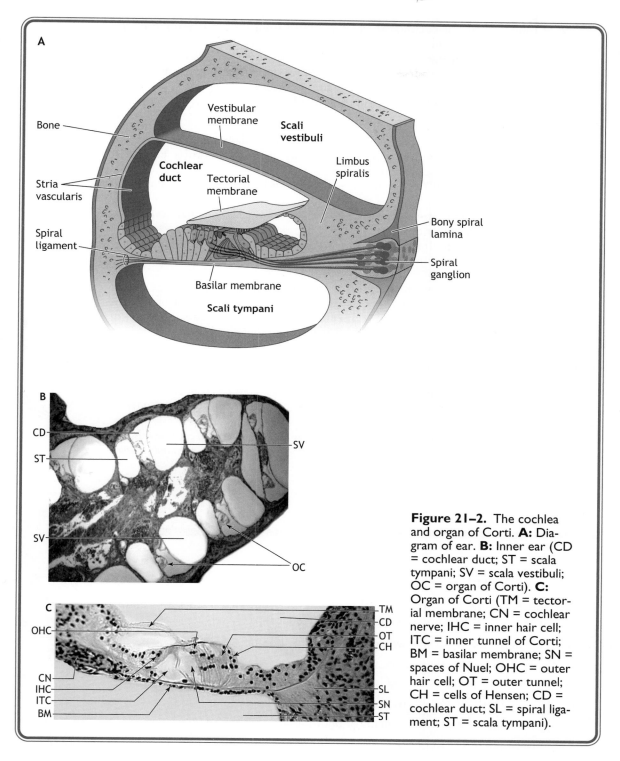

A:

Bone

Vestibular membrane

Scali vestibuli

Cochlear duct

Tectorial membrane

Limbus spiralis

Stria vascularis

Spiral ligament

Bony spiral lamina

Spiral ganglion

Basilar membrane

Scali tympani

B

CD

ST

SV

SV

OC

C

OHC

CN
IHC
ITC
BM

TM
CD
OT
CH

SL
SN
ST

Figure 21–2. The cochlea and organ of Corti. **A:** Diagram of ear. **B:** Inner ear (CD = cochlear duct; ST = scala tympani; SV = scala vestibuli; OC = organ of Corti). **C:** Organ of Corti (TM = tectorial membrane; CN = cochlear nerve; IHC = inner hair cell; ITC = inner tunnel of Corti; BM = basilar membrane; SN = spaces of Nuel; OHC = outer hair cell; OT = outer tunnel; CH = cells of Hensen; CD = cochlear duct; SL = spiral ligament; ST = scala tympani).

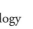

2. A central canal, the **cochlear duct,** is separated from the scala vestibuli by the **vestibular membrane** (Reissner's membrane) and is filled with **endolymph.**

3. The **basilar membrane** separates the scala media from the scala tympani. The basilar membrane extends from the spiral limbus to the **spiral ligament;** the former is attached to the central bony core of the cochlea.

4. The spiral limbus produces the **tectorial membrane.**

E. Organ of Corti

1. The **organ of Corti** (Figure 21–2C) rests on the **basilar membrane,** whereas the **tectorial membrane** overlays the stereocilia of **inner** and **outer hair cells.** These are the sensory cells of the inner ear.

2. The **tunnel of Corti** (inner tunnel) is flanked on either side by the inner and outer pillar cells.

3. There are 3 rows of **outer hair cells** and a single row of **inner hair cells** adjacent to the inner pillar cells.

4. Dendrites extending from the **spiral ganglion cells** pass beneath the spiral limbus and through the inner tunnel of Corti and spiral around the bases of sensory hair cells.

5. As sound waves travel, they distort the **vestibular membrane** and the basilar membrane of the organ of Corti, which, in turn, causes movement of the hair cells in contact with the **tectorial membrane.** This mechanical stimulation of the **hair cells** is transferred to the dendrites of the spiral ganglion cells.

COCHLEAR DISORDER

- *Deafness* can result from damage to the *cochlear nerve,* the vasculature, and bony structures. Sensorineural deafness results from a lesion of the cochlear nerve or a lesion of the cochlea. *Mondini's deafness* results from dysgenesis of the *organ of Corti.*
- Increased bone deposition at the rim of the oval window in the middle ear is called *otosclerosis.* This condition gradually results in bone formation, which anchors the *stapes* at this site, eventually leading to hearing loss.

CLINICAL PROBLEMS

A person suffers a severe blow to the lateral region of the orbit. He presents with blurred vision. You suspect that he has suffered a retinal detachment.

1. At which of the following sites would this detachment occur?

A. Between ganglion cells and the vitreous

B. Between cells of the inner and outer nuclear layers

C. Between the RPE and photoreceptor cells

D. Between inner and outer segments of photoreceptor cells

E. Between photoreceptor cells at the outer limiting membrane

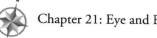

After an eye examination, it was determined that your patient has intraocular pressure exceeding 40 mm Hg. Blockage of aqueous humor flow was suspected.

2. This increased pressure would result from blockage of aqueous humor at which of the following sites?

   A. Choroid body

   B. Vessels at the optic nerve head

   C. Intraretinal vessels

   D. Trabecular meshwork

   E. Choriocapillaris

A patient suffers from elevated intraocular pressure. On fundic examination, a noticeable structural defect was observed at the optic disk. Cupping of the optic nerve head was a possibility.

3. Damage at this site would most directly result in which of the following?

   A. Reduction in blood supply to the retina

   B. Death of ganglion cells

   C. Deformation of the cornea

   D. Loss of vitreous

   E. Increased cells within the vitreous

4. Cell bodies of which of the following cells are found within the inner nuclear layer?

   A. Müller cells

   B. Ganglion cells

   C. Photoreceptor cells

   D. Retinal pigment epithelium

   E. Astrocytes

5. Axons of photoreceptor cells synapse with dendrites of which of the following cells of the retina?

   A. Horizontal cells

   B. Amacrine cells

   C. Ganglion cells

   D. Müller cells

   E. Bipolar cells

6. The organ of Corti lies on which of the following structures?

   A. Spiral limbus

   B. Tectorial membrane

   C. Vestibular membrane

   D. Basilar membrane

   E. Spiral ligament

7. Which of the following is the thickest component of the cornea?
   A. Corneal endothelium
   B. Stroma
   C. Descemet's membrane
   D. Bowman's membrane
   E. Corneal epithelium

A 50-year-old man suffered hearing loss over a period several years. After a series of tests, nerve damage was eliminated as a cause of the loss of hearing.

8. Which of the following structures of the organ of Corti would you suspect played the most significant role in this hearing loss?
   A. Inner pillar cells
   B. Basilar membrane
   C. Vestibular membrane
   D. Inner hair cells
   E. Outer phalangeal cells

A 60-year-old man has worked for more than 20 years in a factory in which machinery produces intermittent loud noises. Remarkably, he has suffered little hearing loss over these years.

9. Which of the following components of the ear prevented these noises from damaging the inner ear?
   A. Vestibular membrane
   B. Helicotrema
   C. Tympanic membrane
   D. Reissner's membrane
   E. Stapedius muscle

A 75-year-old woman has suffered from progressive vision loss over several years. Her sight is extremely limited at night.

10. Which of the following cell types is defective or lost in this patient?
    A. Cone photoreceptor cells
    B. Rod photoreceptor cells
    C. Müller cells
    D. Retinal pigment epithelium
    E. Astrocytes

A 60-year-old woman experienced gradual vision loss that began with difficulty in reading and progressed to blurred central vision. Her peripheral vision was not affected.

11. Which of the following is the proper diagnosis?
    A. Macular degeneration
    B. Retinitis pigmentosa

C. Cataract

   D. Uveitis

   E. Retinal detachment

12. Unmyelinated axons exit the retina to form the optic nerve. Which of the following structures do these axons pass to enter the optic nerve?

   A. Choriocapillaris

   B. Trabecular meshwork

   C. Lamina cribrosa

   D. Macula lutea

   E. Ora serrata

13. The scala tympani and scala vestibuli within the inner ear communicate via which of the following?

   A. Oval window

   B. Round window

   C. Cochlear duct

   D. Eustacian tube

   E. Helicotrema

ANSWERS

1. The answer is C. Retinal detachment occurs between the retinal pigment epithelium (RPE) and the photoreceptor cells. If reattachment is not performed quickly, focal photoreceptor cell death can occur.

2. The answer is D. Aqueous humor produced by the choroid body passes from the posterior chamber to the anterior chamber and then through the trabecular meshwork. Aqueous humor drainage blocked at this site can result in increased intraocular pressure, leading to conditions in glaucoma.

3. The answer is B. Cupping at the optic nerve results in damage to the optic nerve, specifically the axons of ganglion cells. Damage to these axons results in death of ganglion cells by apoptosis.

4. The answer is A. Cells in the inner nuclear layer include not only Müller cells but also bipolar neurons and amacrine and horizontal cells. Astrocytes are found surrounding blood vessels at the vitreal border.

5. The answer is E. Axons of photoreceptor cells primarily synapse with the dendrites of bipolar neurons, which, in turn, synapse via their axons with ganglion cell dendrites. Horizontal cells may have limited synaptic contacts with photoreceptor cell axons.

6. The answer is D. The organ of Corti of the cochlea lies on the basilar membrane. The tectorial membrane lies on the apical segment of outer and inner hair cells. The vestibular membrane separates the scala media from the scala vestibuli.

7. The answer is B. The thickest segment of the cornea is the stroma, making up more than 90% of its thickness. All other components are membranes, a single cell layer, or a stratified cell layer.

8. The answer is D. Inner hair cells and outer hair cells are the sensory components of the organ of Corti.

9. The answer is E. The stapedius muscle is attached to the stapes, which limits the vibration of this bone caused by loud noises.

10. The answer is B. Rod photoreceptor cells are significantly more sensitive to light than cones; thus, rods are responsible for night vision.

11. The answer is A. Central vision loss is a consequence of age-related macular degeneration. Cone photoreceptor cells are concentrated at the macula and are responsible for fine vision used in reading.

12. The answer is C. Axons from ganglion cells exit the retina at the optic disk. These axons pass through the lamina cribrosa, which is a component of the sclera.

13. The answer is E. The helicotrema provides communication of the scala tympani and scala vestibuli at the apex of the cochlea. These channels are filled with perilymph.

INDEX

Page numbers followed by italic *f* or *t* indicate figures or tables, respectively.